Communications under the Seas

Dibner Institute Studies in the History of Science and Technology

Dibner Institute Studies in the History of Science and Technology
George Smith, general editor

Bernadette Bensaude-Vincent and William R. Newman, editors, *The Artificial and the Natural: An Evolving Polarity*

Jed Z. Buchwald and I. Bernard Cohen, editors, *Isaac Newton's Natural Philosophy*

Jed Z. Buchwald and Andrew Warwick, editors, *Histories of the Electron: The Birth of Microphysics*

Geoffrey Cantor and Sally Shuttleworth, editors, *Science Serialized: Representations of the Sciences in Nineteenth-Century Periodicals*

Bernard Finn and Daqing Yang, editors, *Communications under the Seas: The Evolving Cable Network and Its Implications*

Michael Friedman and Alfred Nordmann, editors, *The Kantian Legacy in Nineteenth-Century Science*

Anthony Grafton and Nancy Siraisi, editors, *Natural Particulars: Nature and the Disciplines in Renaissance Europe*

J. P. Hogendijk and A. I. Sabra, editors, *The Enterprise of Science in Islam: New Perspectives*

Frederic L. Holmes and Trevor H. Levere, editors, *Instruments and Experimentation in the History of Chemistry*

Agatha C. Hughes and Thomas P. Hughes, editors, *Systems, Experts, and Computers: The Systems Approach in Management and Engineering, World War II and After*

Manfred D. Laubichler and Jane Maienschein, editors, *From Embryology to Evo-Devo*

Brett D. Steele and Tamera Dorland, editors, *The Heirs of Archimedes: Science and the Art of War through the Age of Enlightenment*

N. L. Swerdlow, editor, *Ancient Astronomy and Celestial Divination*

Communications under the Seas

The Evolving Cable Network and Its Implications

---

edited by Bernard Finn and Daqing Yang

The MIT Press
Cambridge, Massachusetts
London, England

For information about special quantity discounts, please email special_sales@mitpress.mit.edu

This book was set in Bembo by The MIT Press.
Printed and bound in the United States of America.

Library of Congress Cataloging-in-Publication Data

Communications under the seas : the evolving cable network and its implications / edited by Bernard Finn and Daqing Yang.
p. cm. — (Dibner Institute studies in the history of science and technology)
Includes bibliographical references and index.
ISBN 978-0-262-01286-7 (hardcover : alk. paper)
1. Cables, Submarine—History. 2. Telecommunication—Social aspects—History.
3. Communication, International. I. Finn, Bernard S., 1932– II. Yang, Daqing, 1964–
TK5103.15.C66 2009
621.387'8409162—dc22

2008042011

10 9 8 7 6 5 4 3 2 1

Contents

Contributors

Jorma Ahvenainen Professor of History, Emeritus, Jyvaskyla University

Robert Boyce Senior Lecturer in International History, London School of Economics and Political Science

Bernard Finn Curator of Electrical Collections, Emeritus, National Museum of American History, Smithsonian Institution

Pascal Griset Centre de Recherche en Histoire de l'Innovation, Université Paris-Sorbonne

Daniel R. Headrick Professor Emeritus of Social Science and History, Roosevelt University

Jeff Hecht Freelance science and technology writer, Boston correspondent: *New Scientist* magazine, Contributing editor: *Laser Focus World*

Peter J. Hugill Professor in the Department of Geography and in International Affairs Program of the Bush School of Government and Public Service, Texas A&M University

Kurt Jacobsen Professor and Director of the Centre for Business History, Copenhagen Business School

David Paull Nickles Historian, U.S. Department of State

Jonathan Reed Winkler Assistant Professor, Department of History, Wright State University

Daqing Yang Associate Professor of History and International Affairs, George Washington University

The Technological Challenge

1

Introduction

Bernard Finn and Daqing Yang

On December 26, 2006, a 7.1 magnitude earthquake took place under the ocean off Taiwan and caused widespread disruption to telephone and Internet connections in East and Southeast Asia. A little over a year later, in early 2008, damage to cables near Alexandria and elsewhere in the Middle East—its causes still a mystery—affected over 80 million Internet users in the region and in locations like India and Pakistan. Although rerouting helped reduce the adverse impact on business and personal activities to some extent, such incidents reminded the world that, for all the excitement about new technologies such as Wi-Fi, mobile networks, and communication satellites, the bulk of its long-distance communication remains dependent on what one newspaper called "altogether older technology: cables under the earth's oceans." As the public realized that full repairs to these damaged cables would take weeks, if not months, another newspaper remarked that "workers are relying on 19th century technology to fix a 21st century problem."[1]

In the closing decade of the twentieth century we were presented with a worldwide communications system of breathtaking speed and capacity. Economically it was spawned by society's voracious appetite for information, especially as generated by computers and fed over the Internet. Technologically it was fueled by low-loss optical fibers, an indication of the increasing importance of science in an industry that for its first century paid little attention to research.

The implications for society of these advances in communications speed and capacity are difficult to predict. Will they bring greater stability to economic markets or make them more erratic? Will they encourage the expression of multiple voices or the dominance of a few? Will their net effect be to serve the cause of peace or of strife? And will the conflict between wired and wireless be settled—at least for the twenty-first century—in favor of the wires, or will the ether (presumably with the assistance of satellites) mount an effective response? If we cannot predict how these issues will be played out in the decades ahead, we can at least examine how they have been dealt with in the past.

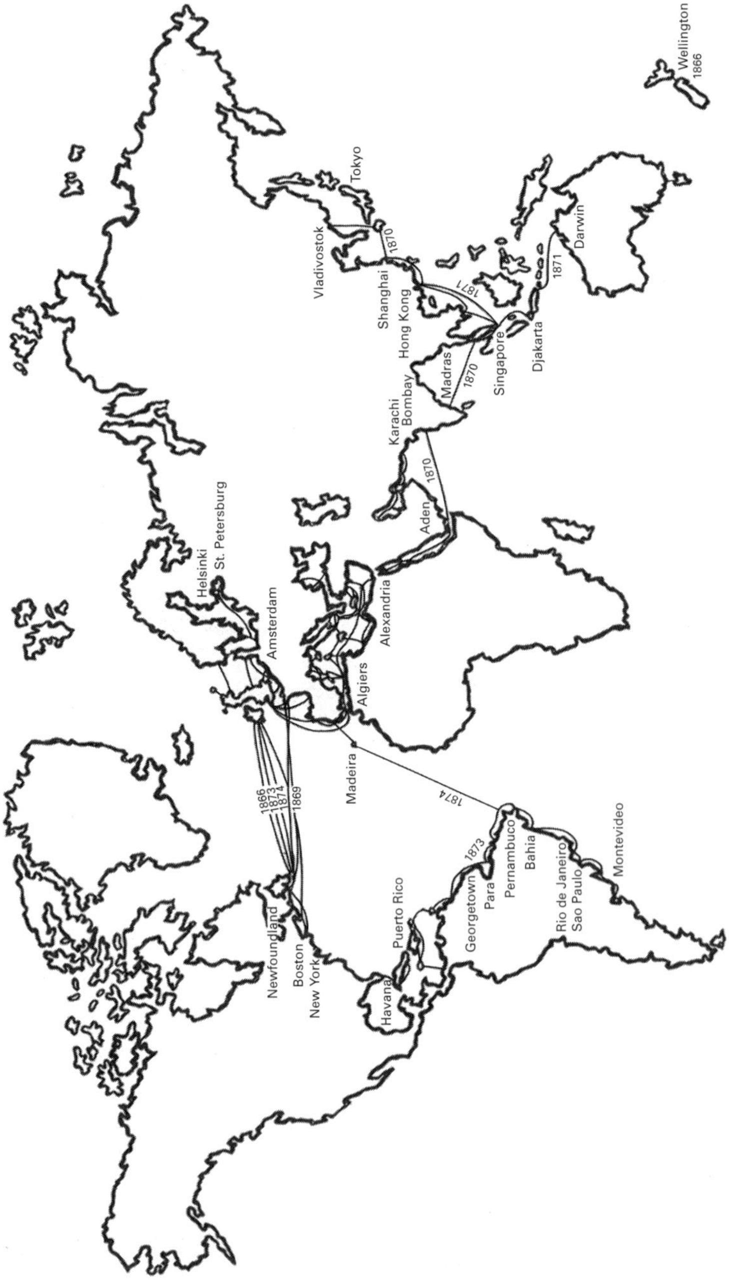

Telegraph cables, 1875.

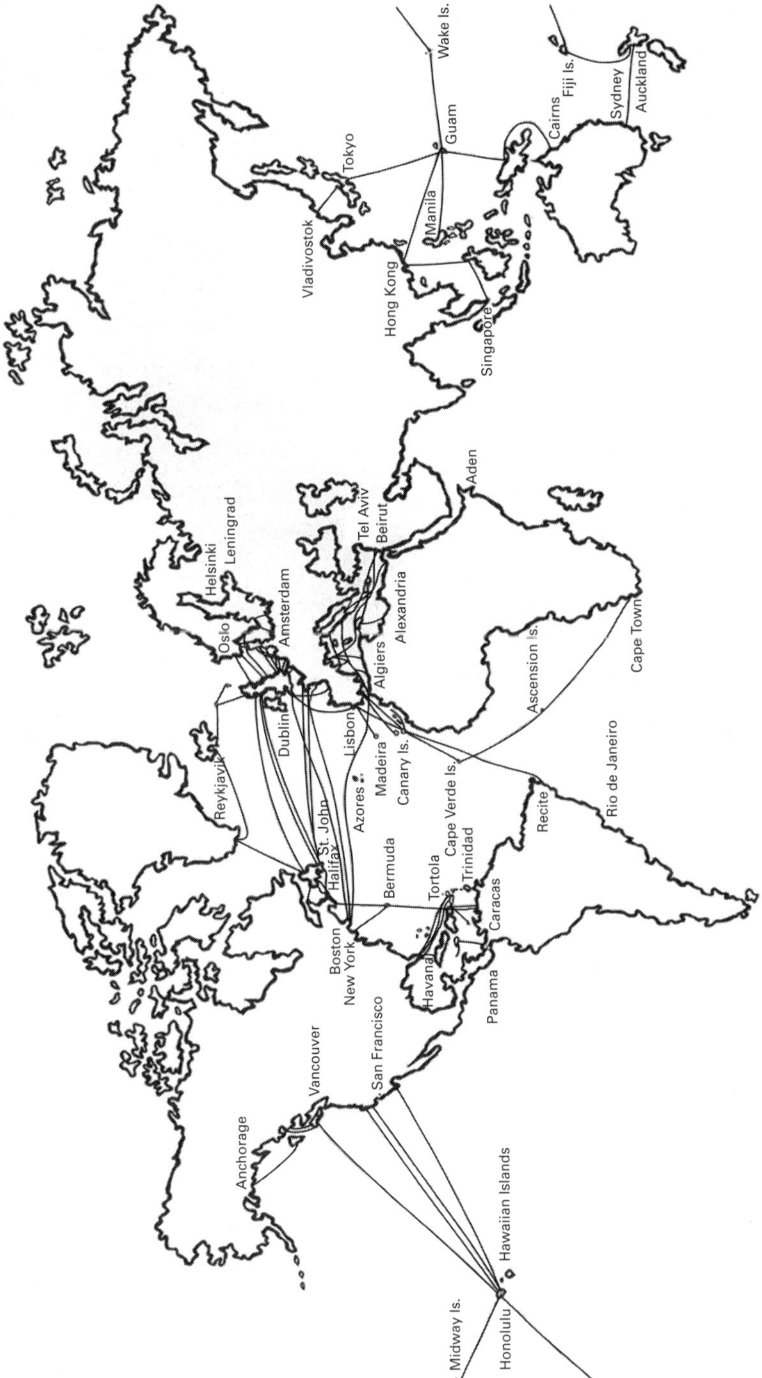

Telephone cables, 1975.

This collection of essays focuses on point-to-point communications (as opposed to broadcasting) and, in particular, on undersea cables. For more than a half century, beginning in the 1860s, submarine electric telegraph cables provided the only means of conveying information in "real time" across large bodies of water. From the beginning of the twentieth century to the mid-1950s this technology was challenged by wireless, especially shortwave radio, which reduced the fixed cost and was capable of transmitting voice. In the 1960s, the emergence of communications satellites seemed to spell the end of intercontinental submarine cables. Since then, however, technological innovations, involving first the introduction of submerged amplifiers and then the substitution of optical fibers for copper conductors, have restored hard-wired systems as the dominant technology for the global network—both figuratively and literally a world wide web.

From the beginning—usually identified with the failed transatlantic cable of 1858 (promoted by American Cyrus Field and built with British capital and technology)—the sheer scope of this evolving enterprise has attracted attention, and several popular accounts have dramatized the challenges involved.[2] Only in recent years, however, have historians looked seriously at the development of the technology itself or at its social implications. In April 2002 some of those most prominently engaged in such studies came together for a two-day symposium sponsored by the Dibner Institute for the History of Science and Technology in Cambridge, Massachusetts. The presentations and discussions were suffused by a special sense that this topic was timely: cable enterprises were international in their historical unfoldings as well as their physical manifestations, and as such they were the harbingers of globalization. Fittingly, the conference had brought together an international group of scholars who were testing the extent of this field of inquiry and laying the groundwork for future studies. In this volume we have the substance of most of those presentations, with modifications inspired by discussion and further thought.[3]

These accounts focus principally on the period from the 1850s to the 1950s, when cables had limited capacities and could not carry voice signals. It seems important, however, to consider technical developments up to the present era, and this is done by the first three papers in this book. Bernard Finn concentrates on the early years, when a financially and technically daring group of entrepreneurs wove a network of insulated copper wires around the globe and then quickly became conservative as they consolidated their gains. Jonathan Winkler takes us from the period immediately following the First World War, when the cables were threatened with economic extinction by a rival international wireless network, to the era following the Second

World War, when new electronic techniques reinvigorated the industry. Jeff Hecht then tells how, in the closing years of the twentieth century, new techniques employing light-carrying glass fibers provided the enormous increase in information capacity required in the Internet age.

The next four papers explore the ways in which governments and private companies interacted in the process of establishing a viable communications network with truly global reach. By the early 1870s the world's major population centers were linked by wire, and a fully reliable system, with sufficient redundancies, was in place by the first decade of the twentieth century. Jorma Ahvenainen provides an overarching account of how government and business forces overcame political, economic, and technical obstacles to make the telegraphic network available to the general public. Examining the crisis of the 1920s, when the cable industry was threatened by cheaper shortwave radio technology, Robert Boyce shows how government and commercial interests in Great Britain successfully forced a merger of wireless and cable companies. Kurt Jacobsen shows how a small company in a small nation (Denmark) used clever tactics based on its nonthreatening position to construct a significant link—across Russia to Japan and China—in the global network. And Pascal Griset examines transatlantic cables from a French viewpoint, describing how negotiations with other governments (Britain and the United States) and corporations (principally the Commercial Cable Company) were complicated by cultural and economic differences.

Three studies look in the other direction, examining the role of cable communication in government activities. David Nickles comments broadly on ways that cables affected the style and effectiveness of the foreign service. Daniel Headrick describes how British dominance of the cable network affected that nation's international strategies and how well the cables served the country during the world wars. Daqing Yang explains how modern Japan's expansion of telegraph cables and its imperialist ambitions became intertwined.

Finally, by way of summary, Peter Hugill places these contributions in the context of concepts of macrotechnology and national hegemonies as developed by Wallerstein, Hall and Preston, Mahan, Mackinder, and himself. As a finale he offers a set of alternative predictions for the decades ahead.

This book is not intended to be the final word. By focusing attention on how undersea communication networks have shaped and been shaped by government institutions, affected by technological change and stagnation, we hope to stimulate further investigations. One might, for instance, expand our coverage to include other countries or regions. But it would be especially interesting to look at the relationships of the evolving cable networks to other elements of society: financial, corporate, political, and cultural, as well as personal. Communication is at the heart of them all; thus the quality, cost, speed, accessibility, and stability of communications is inevitably of great significance. Such historical investigations will likely yield a better appreciation of the potential and the fault lines of our communication-dependent global society.

Notes

1. "Damaged Phone Cables Get Low-Technology Fix," *Taipei Times*, January 13, 2007; "Underwater Peril," *Business Week*, January 15, 2007; "2008 Submarine Cable Disruption," Wikipedia (accessed November 22, 2008).

2. The first popular account was the sixteen-page "Telegraph Supplement" published by *Harpers Weekly* on September 4, 1858—just as the Field cable was being declared dead. William Howard Russell wrote a more extensive narrative, *The Transatlantic Submarine Telegraph* (London: Dawson, 1866), with color illustrations by Robert Dudley. These were followed by Henry M. Field (Cyrus's brother), *The Story of the Atlantic Telegraph* (New York: Scribner's Sons, 1892), and Charles Bright (son of the engineer who oversaw the laying of the first Atlantic cables), *The Story of the Atlantic Cable* (New York: Appleton, 1903). The centennial of the first Atlantic cable inspired works by Arthur C. Clarke, *Voice across the Sea* (New York: Harper, 1958), and Bern Dibner, *The Atlantic Cable* (Norwalk: Burndy Library, 1959).

3. In addition to the those contributing to this volume, Menahem Blondheim of the Hebrew University in Jerusalem and Janet Abbate, then of the University of Maryland, made valuable formal presentations at the conference. Among others in attendance, John Britton of Francis Marion University and José Altshuler of the Sociedad Cubana de Historia de la Ciencias y la Tecnología were especially helpful in sharing their observations with the conferees.

2

Submarine Telegraphy: A Study in Technical Stagnation

Bernard Finn

Recent studies of technical change have dealt with different aspects of what is commonly recognized as a complex process. Even historians who focus narrowly on the act of invention—on the solution of a particular technical problem—emphasize the range of issues confronting the inventor. For those who include in their view the invention's introduction into the marketplace and its eventual effects, the multitude of significant factors can become very large indeed. At this level some historians see a primacy in social forces shaping a relatively pliant set of technological givens. For others, successful innovation is seen as resulting from solutions to a succession of problems (technical, economic, social) that constitute a system. It is not my purpose to project a new theoretical approach or even to introduce a new set of terms to the litany of "reverse salients," "technological frames," and "communities of practitioners." But because most, if not all, case studies examined in these analyses have been examples of successful innovation or (less frequently) of failure, it may be fruitful to consider an example of relative stagnation. In this example a technology survived basically unchanged for a hundred years. I shall suggest reasons for its resistance to change and for why, in spite of that resistance, it survived.

Submarine telegraphy was a significant innovation, even if those who were responsible did not see themselves as doing anything remarkable from a technological point of view. When John and Jacob Brett proposed in 1845 that a fledgling landline telegraph system should be extended across the English Channel (and indeed to the rest of the world), they considered their main problem to be political—persuading the French and English governments to grant them the necessary concessions.[1] And a decade later, when Cyrus Field was offered the opportunity to take over a bankrupt enterprise attempting to construct an overland telegraph from Boston to St. Johns, Newfoundland, it took him but a moment to look at the globe in his study and decide to go the rest of the way—to Ireland. He thought that his main problem would be raising money.[2]

PUNCH, OR THE LONDON CHARIVARI, AUGUST 21, 1858.

THE ATLANTIC TELEGRAPH—A BAD LOOK OUT FOR DESPOTISM.

JOHN BULL. "HOLD FAST, JONATHAN." JONATHAN. "ALL RIGHT, JOHNNY."

Cartoon from *Punch* (August 21, 1858).

"NEARER AND DEARER."—THE SUBATLANTIC SPLICE.

Cartoon from *Punch* 35 (1858), 73.

But of course it was much more than money. The technology of underwater cables turned out to differ significantly from that of landlines. Some of this was apparent when the Bretts crossed the Channel in 1850; and it would be painfully obvious when Field and his British colleagues bridged the Atlantic in 1858. Among the more important problems were the following:

• Since most cables were international, there were often complex political negotiations to be pursued as well as agreements to be made with local telegraph companies regarding fees and technical standards.

• Promoters needed steamships to carry many miles of heavy cable; the ships had to be equipped with apparatus for laying (and occasionally retrieving) their cargo.

• The copper core of the cable itself had to be electrically insulated from the water and also physically protected from tides and anchors and other potential sources of abrasion.

• The physics of what happened to an electrical signal as it tried to force its way through what might be described as a long, thin capacitor had to be understood.

• Markets would have to be established that were large enough to justify the substantial capital investment.

Suffice it to say that the basic elements of cable technology were established early and remained largely unchanged for a century.

CABLE TECHNOLOGY

The 25-mile cable that the Brett brothers laid from England to France consisted of a solid copper wire coated with one-half inch of gutta-percha. It was manufactured in 100-yard lengths, which were then joined together. Almost as an afterthought, lead weights were added to keep this lightweight assemblage firmly at the bottom. The Bretts realized that the shore ends would need additional protection, so here the wire was coated with cotton soaked in an india rubber solution and encased in a lead tube. The cable was successfully laid, but it failed after a few hours, when a fisherman caught and broke it.[3]

A new cable, laid in 1851, had four separate conductors, each covered with two layers of gutta-percha, embedded in tarred hemp and protected with an outer covering of iron wire. At about seven tons per mile, the weight of the cable had now become significant, for it was difficult to control the rate at which it was payed out over the stern of the boat. In fact, the end was reached a mile short of shore. Within a month a new section had been produced, and on October 19 the four lines were open to public use.[4]

A pattern had been set that was followed by other early cables—to the Netherlands and Ireland, from Denmark to Sweden, and in the Mediterranean. In 1856 the line between Nova Scotia and Newfoundland, a section of the Field line, was given a conductor of seven strands of wire wound together. This design, which assured continuity of service even if one of the wires broke, immediately became standard practice.[5] In subsequent years, as cables extended south and east into warmer seas, they met with insects that had a propensity for boring into the gutta-percha. This problem was solved with addition of a copper or brass sheathing, which was first employed in 1879 and became a common addition.[6]

The most problematic cable constituent was the insulator, gutta-percha. It had appeared as a miracle substance, just in time for its God-appointed use. The gum from a tree that grows in Malaysia, primarily on the Malay Peninsula and on the islands of Java and Borneo, gutta-percha was introduced to European science and technology circles in 1843. Its excellent insulating properties were determined by Werner Siemens in 1846 and by Michael Faraday in 1848. It is a natural plastic, firm at ordinary temperatures but becoming malleable when the temperature approaches 40 degrees centigrade. It can become brittle when exposed to air but is virtually inert under water, especially in the temperature and pressure conditions at the bottoms of the world's oceans. Very little needed to be done to it, and techniques were quickly developed for processing the imported boules of gutta-percha so that it could be extruded around copper wire cores.[7]

No one seemed to worry about the supply. A typical cable contained about 400 pounds of gutta-percha per mile (some light cables contained a little less than half that amount), meaning that very large quantities were soon needed. In 1881 alone some 3,600 tons were exported from Singapore,[8] and over the course of the last half of the nineteenth century tens of millions of trees were cut down for their sap (which does not flow as easily as rubber). Finally, in 1915 the British company that dominated the field (Telegraph Construction and Maintenance Company) established a plantation north of Singapore in Malaya (the Dutch had preceded them with one in Java). They used a less destructive harvesting technique, relying on leaves and twigs.[9] Some relief was also found in a species of tree in South America that produced balata, a cousin of gutta-percha that was more resinous and therefore required more treatment. Eventually polyethylene would replace gutta-percha, but this discovery of the 1930s would not be applied to cables until the 1940s.

A major development came in the mid-1920s with the invention of Permalloy, an iron-nickel magnetic material. It could be wrapped around the core of the cable to provide "inductive loading" that helped to offset the capacitance effect and to increase substantially the volume of messages that could be sent. But the critical work on this was done at Western Electric (and taken over by Bell Laboratories when it began operations in 1925), not Western Union. Mu-metal, developed subsequently at TC&M, was effectively an adaptation of the Bell process.[10]

Manufacturing

Within a very short time, basic manufacturing techniques were established for processing gutta-percha and extruding it around the stranded copper wire and for winding hemp and iron wires around the core. One can only marvel at the ability of the nascent cable industry, in 1857, to produce 2,500 miles of cable for the Field enterprise in four months. As Charles Bright wrote:

> To give an idea of what this meant, it will suffice to say that no less than 119½ tons of copper had to be provided for the construction alone. The copper had to be drawn out into 20,500 miles of wire (providing for the lay); and seven parts of this wire had to be laid up in a strand 2,500 miles long. For insulation, moreover, nearly 300 tons of gutta-percha were required to be prepared, and applied to the conductors in three separate coatings. Lastly, and with due allowance for lay, 367,500 miles of wire had to be drawn from 1,687 tons of charcoal iron, and laid up into about 50,000 miles of strand for the outer sheathing.[11]

As Bright was quick to point out, this frantic pace was an invitation for trouble—not for the 1857 attempt, when only a small portion was used, but for 1858. For example, there was no time to construct water tanks to store the cable as it was manufactured. It was coiled up in the open air on the wharf, where the sun's heat could soften the gutta-percha, allowing the copper core to shift and come close to or even touch the iron sheathing wires. The pressure of time also undoubtedly contributed to a degree of sloppiness in the manufacturing process.

As we know, the 1858 cable lasted for barely a month and was never open for commercial traffic. Its failure was probably caused by the high-intensity electrical pulses at first employed by the electrician, breaking down the resistance of the insulation in a cable made vulnerable by the problems cited above. A year later more cable (though somewhat lighter) was manufactured for a line down the Red Sea and across to India. Although the manufacture was presumably at a more relaxed pace, the laying was not. Influenced (Bright suggests) by a condition in the contract that gave any excess cable to those who were laying it, the engineers in charge allowed for very little slack, with the result that the cable was suspended between ridges on the uneven bottom, and it quickly broke under its own weight.[12]

The British government had a significant stake in both cables, expressed in the form of traffic guarantees. The twin disasters stimulated Parliament to establish a joint committee to investigate the failures. In 1861 the report of the committee pointed its finger at testing and laying procedures and at purity of materials, not at the manufacturing processes.[13] The failures were consequences of the overexuberance of promoters, who had little knowledge of the technology and were more interested in getting the job done than in doing it well. Even before the report was issued, the engineers were in fact getting it right with a series of cables that Charles Tilston Bright laid in the Mediterranean. After Bright's cables, there were virtually no reported problems associated with faulty cable design and construction.

The principal difficulty in laying the first cables was controlling the speed at which the cable went over the stern of the ship. (The Red Sea cable, however, as we have seen, was let out in relatively shallow water too slowly rather than too fast.) This technical challenge was clearly going to be a major concern in the deep water of the Atlantic. In the small amount of time available to him in the spring of 1857, Bright devised a system that guided the cable around four drums; a dynamometer measured tension, and a braking mechanism slowed the payout if necessary. As already mentioned, the system was not entirely satisfactory, contributing to a fatal break that occurred when barely 300 miles had been laid. The following year Bright completely rearranged the components and created a system that was copied well into the twentieth century.[14]

Amazingly, the 1857 and 1858 expeditions were undertaken without any equipment capable of picking up cable from the ocean floor, in the very likely event that it should break. In 1865 the question had still not been addressed. When a break did occur, two-thirds of the way across the Atlantic, only makeshift equipment was available for an attempted recovery (including buoys, anchor wire and rope). This led to a series of experiments in the Bay of Biscay, together with detailed mathematical analysis by William Thomson (later Lord Kelvin),[15] and in the summer of 1866 the expedition was fully prepared, both with equipment and training, to recover the 1865 cable. Expeditions from that point on would be similarly prepared.

INSTRUMENTATION

None of the early promoters or engineers understood the effect of laying cable underwater. They moved blindly ahead, believing that landline techniques and equipment would suffice. For short cables they were right. But to their dismay (both psychological and financial) they found in 1858 that this was not true for long cables. Again it was William Thomson who provided the critical information. He had already analyzed the propagation of signals in cables, showing that the capacitance effect produced by the surrounding water tends to smear out the signal.[16] As a consequence, a sharp input signal is not at first detected at the receiving end. The signal is received as a bell-shaped curve, rising slowly to a peak and then trailing off. The practical consequences of this, as perceived by Thomson, were that a sensitive receiver was much more important than a powerful transmitter.

So Thomson designed his mirror galvanometer,[17] which was a modification of the tangent galvanometer described by Hermann von Helmholtz in 1853.[18] A lightweight mirror carrying light magnets on its back was suspended in the center of a coil. The signal current flowing one way in the coil caused the mirror to twist, an effect that could be seen by the motion of a light beam reflected from the mirror. If the signal current flowed the other way, the mirror would twist in the opposite direction. This instrument allowed operators to receive a few messages for about a month before the cable went dead. The mirror galvanometer was used exclusively until the early 1870s, and it continued to be used for very sensitive testing purposes until the demise of the unrepeatered cables in the 1950s.

In 1867 Thomson patented his siphon recorder. With improvements, it was introduced in the Hearts Content station in 1874. In this instrument a lightweight coil was suspended between the poles of a powerful magnet. Signal current from the cable caused the coil to twist (one direction for a "dot" and the other for a "dash"). Threads connected the coil to a thin glass tube that siphoned ink from a small inkwell to a position close to but not touching a moving paper strip. The threads moved the siphon back and forth. The ink, which was electrically charged, spurted onto the paper in wavy lines, which could then be interpreted as coded letters. The siphon recorder was just about as sensitive as the mirror galvanometer, and it had the great advantage of producing a permanent record. As a result, in spite of the cost (augmented substantially by royalties during the term of the patent), the siphon recorder became the receiver of choice for most long cables, and it remained in use until the end.[19]

In the twentieth century several instruments were devised that would serve as amplifiers, relaying (and shaping) the signal to a receiver—often a siphon recorder—or even automatically retransmitting it to the next station. All of these instruments were based on Thomson's principle of threads attached to a moving coil. There were reasons for this. As two veteran British telegraph engineers noted in 1932,

> When one studies the succession of receiving instruments that have characterized cable progress, the remarkable persistence of the d'Arsonval galvanometer is noticeable. The siphon recorder was, due to Lord Kelvin's genius, so successful that nearly every other receiving instrument has the elements of the original recorder, or the galvanometer on which it was based, its moving system. . . . One reason is that such apparatus possesses mechanical selectivity which effectively eliminates extraneous interference of frequencies above that of the working range, and another is that its sensitivity is sufficient for the finest duplex balances and lowest interference levels attainable in practice. . . . [And except for the newest cables, their adoption] has been prevented by the consideration that no higher return in traffic capacity has been promised, and no incentive offered to write off the capital sunk in existing apparatus.[20]

Several of these were developed by British instrument firms in the period 1910–1920. In each case threads from the moving coil either upset the balance of a Wheatstone bridge or changed an electrical contact. In the Heurtley magnifier the temperatures (and therefore the resistances) of two arms of a bridge are affected; in the selenium magnifier a beam of light flips from one selenium cell to another, where the cells are in a bridge circuit; in the Muirhead, Bruce, and Brown relays a contact arm is moved. The Brown model was especially popular with Western Union. The arm rested lightly on drum with two slightly separated conducting strips. Rotation of the drum reduced the frictional force on the arm as it moved back and forth.[21]

Laboratories

Technical innovation was hindered by the fact that cable companies failed to develop experimental laboratories, even after the turn of the century.

On the British side of the Atlantic, the Telegraph Construction and Maintenance Company (TC&M) supported a small operation known as "the Chemistry" (a reflection of an original purpose to study the properties of gutta-percha). Until 1890 this consisted of three people operating in a fashion "clouded in mystery and ignorance, [where] experiments were carried out on a hit or miss system."[22] In that year the activities were placed on a more formal experimental footing, but the focus was on production control.[23] After the war, further expansion brought in trained chemists and physicists. They were therefore ready, in 1921, to see the advantages in the Permalloy material developed at Western Electric and to respond with their similar mu-metal two years later. The available evidence suggests that in other projects research was confined largely if not completely to operations. This was certainly true in America.

Western Union traces its origins back to the formation of the New York and Mississippi Valley Printing Telegraph Company, established in 1851 and using the patents of Royal E. House to challenge Samual Morse interests. Licensing agreements followed, acquisitions were made, and in 1856 the company assumed the name Western Union. More companies were gathered into the fold, and by the 1880s Western Union dominated American telegraphy, with some 90 percent of the business. It was in a strong position, through the patents of Elisha Gray, to challenge the upstart American Telephone and Telegraph Company (AT&T), but in the end the company decided to give way without a serious fight and sold its rights for 20 percent of the profits during the life of the patents. In the current vernacular: big mistake. But Western Union was still a large and important company. In 1890, with gross revenues of $20.1 million, it was slightly ahead of AT&T and comparable in income to Edison General Electric, Westinghouse, and Thomson-Houston combined.

Early in the new century, when these other companies (notably AT&T and General Electric, which represented the joined forces of Edison and Thomson-Houston) established major research laboratories, Western Union did not. The company maintained a traditional shop laboratory as early as the 1860s, and they continued to support the same kind of operation in the new era. In 1900, a space of about four thousand square feet contained the best of machines and apparatus then available for experimental work on telegraph equipment.[24]

About a dozen men were employed in this shop when AT&T purchased control of Western Union in 1909. During the next five years (the government forced a divestiture in 1914) Theodore Vail, president of AT&T, promoted an extensive program of reform within Western Union. Surprisingly, these efforts brought about no significant changes in the shop, in spite of Vail's creation of a Research Branch at AT&T in 1911 (forerunner of Bell Labs). F. E. d'Humy, writing in 1935, noted the Research Branch's focus on improving existing technology:

> Under the circumstances existing during the past twenty-five years or thereabouts, the efforts of the Engineering organization of the telegraph company have necessarily been directed more to the refinement of apparatus, the intensive use of facilities and the development of methods, etc., than to the production of history-making devices or changes in the art.[25]

Instigators of Inertia

In his apology for Western Union's failure to introduce history-making devices, d'Humy speculated that such innovation occurred in technical enterprises that were still "novelties": enterprises that had reached the "stage of stability" were more apt to confine themselves to the "development and improvement of a well-established art." He felt that telegraphy had reached the stage of stability at least by 1909, the telephone had reached it by the time he was writing (1935), and radio was still in the novelty stage.[26] This seems like an unsatisfactory explanation, especially since we know a lot more than d'Humy about what was happening and would happen at Bell Labs. Let me mention some significant factors that were at work.

Entrepreneurial Exuberance

As suggested above, at the beginning the cable industry was pushing the edge of what was technically feasible. Telegraphy itself had barely been established as a viable commercial enterprise—by Cooke and Wheatstone in England in 1842 and by Morse in the United States in 1844. Gutta-percha, the miracle insulator—which equally miraculously turned out to be available in large quantities—had just been "discovered." Iron wire rope manufacture dated back to the 1830s, but a critical patent for winding the wires around a soft (hemp) center was taken out by the British firm R. S. Newall in 1840.[27]

Equally important was the availability of ships to lay the cable. Fortunately, steam engines—an absolute necessity for laying cable in a straight line at constant speed—were replacing sail. But there were few steamships of sufficient size to carry multiple tons of coiled cable. On at least two occasions—for the Sardinia-Bona line in 1854[28] and the link between Nova Scotia and Newfoundland in 1857[29]—sailing vessels were chartered to hold the cable and were towed by steamships. In 1857, when the British battleship HMS *Agamemnon* and the American frigate USS *Niagara* were lent by their respective governments to the Atlantic Telegraph Company, they were two of only six steamships in the world capable of carrying half of the almost 2,000 miles of cable needed to span the Atlantic. In 1859 the *Great Eastern* was placed into service as a passenger liner, only to prove a financial disaster for its owners. It was thus available for the new Atlantic attempts after the American Civil War, and it could carry the full length of cable.

As Bruce Hunt has demonstrated, the uniqueness of the cable enterprise stimulated scientific investigation of a sort not required by landline networks.[30] What is important, however, is that the science was available—as it wasn't just a few years earlier. At the parliamentary committee hearings and in publications, William Thomson was able, on the basis of his pioneering work in mathematical physics of the preceding decade, to provide critical analysis of the effect that capacitance had on the shape of the electrical signal pulse, to emphasize the importance of purity in the copper conductor, to calculate the stress on the cable as it was being laid, and to design appropriate sending and receiving instruments.

For cable entrepreneurs such as papermaker Cyrus Field and textile manufacturer John Pender, this confluence of the essential elements of infrastructure represented simply a happy circumstance. I argue, on the contrary, that because these elements had only *just* become available, there was less likelihood that someone else would be in the wings ready to press these entrepreneurs with a competing system, thus possibly stimulating further innovation. Consider, by way of contrast, Bell's telephone, which could easily have been invented thirty years before its debut and which was immediately contested by several rivals. Patent protection, at least in the United States, preserved the monopoly (which was then maintained for other reasons) but did not prevent the company from encouraging further research. Another example is the incandescent lamp, which waited only for one major technological piece of infrastructure before its debut at the end of 1879. This was the self-excited dynamo, which made substantial amounts of electricity commercially available by the mid-1870s. Numerous rivals to Edison's invention appeared even during the patent honeymoon, and in the twentieth century major research facilities evolved. The core elements of each of these technologies—the carbon transmitter (which quickly replaced Bell's magneto device) and the magneto receiver for the telephone, and the incandescent filament (though tungsten replaced carbon in a major innovation after some thirty years) in an evacuated glass bulb for lighting—lasted even longer than the gutta-percha cables.

Monopoly

Surely one of the strongest forces resisting innovation is lack of competition. Not until the mid-1920s would the cables be challenged by an alternative technology. In the meantime, various factors limited undue competitiveness within the industry.

Most important was the head start enjoyed by the British, stimulated by their unique economic and political reasons for establishing a worldwide network. Except across the Atlantic, the long-line electrical traffic needs of the world were modest and largely satisfied by a single system. The British companies moved quickly to reach the major trade centers around the globe (many of them British colonies), and a basic system was in place by 1872.

This activity in turn brought prosperity to British cable manufacturing companies, which were, in order of importance, the Telegraph Construction and Maintenance Company, Siemens Brothers, and the India Rubber, Gutta Percha and Telegraph Works Company. This inevitably meant that when others—Americans, Greeks, Italians—wanted to lay cables, they turned to one of these companies for its manufacture. Toward the end of the century the French and the Germans became sufficiently concerned about the British position that they financed their own manufacturing operations, but any stimulus for innovation was apparently negligible.

The British also controlled much of the gutta-percha supply. They shared the world's "proven reserves"—in Malaysia (principally the Malay Peninsula, Borneo, Java, Sumatra)—with the Dutch, and Singapore was the port through which virtually all shipments were made.

Across the Atlantic, on the other hand, where traffic demands were heavy, there was room for substantial competition. As early as 1869 a French company sponsored a cable from Brest to the island of St. Pierre (in the mouth of the St. Lawrence) with an extension to Cape Cod. The cable was laid by TC&M, using the *Great Eastern*. Another French cable followed in 1879 (by Siemens Brothers). In both cases a joint-purse arrangement assured that in fact there would be no economically disagreeable contest.[31] This situation changed when silver mining baron John Mackay and *New York Herald* publisher Gordon Bennett formed the Commercial Cable Company. Their initial two cables, in 1883 and 1884, were eventually followed by three more, and they were quite effective in competing for business. Other cables were added to the mix—a total of a dozen by 1900, eighteen in 1920, and a peak of twenty-one in 1928.[32] There is little indication, however, that competition was sufficient to stimulate any major effort in new research.

Alternative Technologies

When real competition—in the form of shortwave radio—came on the scene in the 1920s, the cable industry barely reacted. In Britain the response was political rather than technical: in 1928 the government instigated the merger of the Marconi's Wireless Telegraph Company with the Eastern and Associated Telegraph Companies into (as renamed six years later) Cable and Wireless Ltd.[33] This marriage of convenience had the effect of promoting cooperation instead of competition and almost certainly saved the Eastern companies from bankruptcy. Indeed, with the advent of the Depression, the combined company had a difficult time during the prewar years. In any case, the result was not to stimulate innovation to any great degree in the cable technology.

In the United States, at Western Union, which had been operating the British cables under a lease agreement since 1912, the response was somewhat different:

> Up to 1924 the state of the radio art was such as not to give serious concern, but about this time a radical change took place. Short waves or high frequencies were being rapidly developed and many more channels of communication were opened up. Early in 1925 it was deemed necessary to deal more specifically with the art, and a permanent high frequency laboratory was constructed.[34]

The site selected for the laboratory was on a hill in an isolated section of Long Island, called Water Mill. It was, however belatedly, a classic technical response to competition. It led, in 1948, to a series of 21 microwave towers connecting New York, Washington, Pittsburgh, and Philadelphia.[35] All of this was of course no help to the cable business. And it turned out not to be of particular help to the rest of Western Union because of countermoves by AT&T. In 1950 Western Union produced a truly innovative device: an underwater electronic repeater. It was bulky, and its life expectancy was uncertain. As a consequence, it was connected to cables at the edge of the continental shelf, where it could boost the signal through the final (and noisier) section, and where it could be retrieved for repair with relative ease.[36] But this too was too little, too late. Six years later AT&T and the British General Post Office would lay the first transatlantic fully repeatered telephone cable and make all of Western Union's cables obsolete.

A Culture of Lassitude

Most difficult to define, but perhaps most important, was a corporate culture that was either content to let things be quiescent or that was inept in its ability to promote meaningful change. Such a culture can sometimes be overcome by a strong leader—a prime example is Theodore Vail at AT&T in the early years of the twentieth century. But without forceful direction, corporate inertia can resist even mortal external challenges.

I downplayed d'Humy's "stage of stability," mentioned above, but in at least one sense this characterization is a useful generalization. If a company has a stable workforce and if there are few stimuli for new ideas, then some lethargy takes hold and change becomes more difficult.

When Western Union constructed a new central laboratory facility in 1916, employing fifteen people, there was, it would seem, every opportunity for new explorations. Indeed, the number of patent applications, which was averaging two per year, jumped to eight per year in the 1920s. But the inventions were overwhelmingly operations-oriented: pole preservatives and means of correcting for the effects of earth currents on cables. Few of the staff joined professional societies, and even fewer contributed papers to the meetings. (In 1926 in New York City, 25 Western Union employees were members of the American Institute of Electrical Engineers, compared to AT&T's 367.) An outlet for their research, the *Western Union Technical Review* didn't appear until after the Second World War.

Notes

1. John W. Brett, *On the Origin and Progress of the Oceanic Electric Telegraph* (London: W. S. Johnson, 1858), iii.

2. Henry M. Field, *The Story of the Atlantic Telegraph* (New York: Scribner's Sons, 1892), 16.

3. Charles Bright, *Submarine Telegraphs: Their History, Construction, and Working; Founded in Part on Wünschendorff's "Traité de Télegraphie Sous-Marine" and Compiled from Authoritative and Exclusive Sources* (London: Crosby Lockwood, 1898), 5–9.

4. Bright, *Submarine Telegraphs*, 10–12.

5. Ibid., 27.

6. Ibid., 384.

7. Th. Seeligmann, *Indiarubber and Gutta Percha: A Complete Practical Treatise on Indiarubber and Gutta Percha in Their Historical, Botanical, Arboricultural, Mechanical, Chemical, and Electrical Aspects* (London: Scott, Greenwood, 1903).

8. Bright, *Submarine Telegraphs*, 258.

9. G. L. Lawford and L. R. Nicholson, *The Telcon Story, 1850–1950* (London: Telegraph Construction and Maintenance Co., 1950).

10. M. D. Fagen, ed., *A History of Engineering and Science in the Bell System: The Early Years (1875–1925)* (Murray Hill, N.J.: Bell Telephone Laboratories, 1975), 801–814; Lawford and Nicholson, *The Telcon Story*, 93–98.

11. Bright, *Submarine Telegraphs*, 34–35.

12. Ibid., 57.

13. *Report of the Joint Committee Appointed by the Lords of the Committee of Privy Council for Trade and the Atlantic Telegraph Company to Inquire into the Construction of Submarine Telegraph Cables: Together with the Minutes of Evidence and Appendix Presented to Both Houses of Parliament by Command of Her Majesty* (London: Printed by George Edward Eyre and William Spottiswoode for Her Majesty's Stationery Office, 1861).

14. Bright, *Submarine Telegraphs*, 37, 41–43.

15. William Thomson, "On the Forces Concerned in the Laying and Lifting of Deep-Sea Cables," *Proceedings of the Royal Society of Edinburgh* 5 (1866): 495–509.

16. William Thomson, "On the Theory of the Electric Telegraph," *Proceedings of the Royal Society of London* 7 (1855): 382–399; William Thomson, "On Practical Methods for Rapid Signalling by the Electric Telegraph," *Proceedings of the Royal Society of London* 8 (1856): 299–307.

17. Thomson, "On Practical Methods for Rapid Signalling."

18. Hermann von Helmholtz, "Über einige Gesetze der Vertheilung elektrischer Ströme in körperlichen Leitern, mit Anwendung auf die thierischen elektrischen Versuche," *Annalen der Physik* 89 (1853): 359.

19. For a detailed description, see Bright, *Submarine Telegraphs*, 605–622.

20. H. Kingsbury and R. A. Goodman, "Methods and Equipment in Cable Telegraphy," *Journal of the American Institute of Electrical Engineers* 70 (1932): 488.

21. H. H. Haglund, "Atlantic Telegraph Cable Centennial," *Western Union Technical Review* 12 (1958): 86–88.

22. Lawford and Nicholson, *The Telcon Story*, 93.

23. Ibid., 94.

24. F. E. d'Humy, *A Brief Outline of the Technical Progress Made by the Western Union Telegraph Company, 1910–1934*, limited edition, circulated in typescript (Western Union Engineering Department, 1935), 29.

25. Ibid., 8.

26. Ibid., 1, 7–8.

27. K. R. Haigh, *Cableships and Submarine Cables* (London: Adlard Coles, 1968), 34.

28. Ibid., 29.

29. Ibid., 30.

30. Bruce Hunt, "Michael Faraday, Cable Telegraphy and the Rise of Field Theory," *History of Technology* 13 (1991): 1–19.

31. Samuel Carter, *Cyrus Field: Man of Two Worlds* (New York: Putnam, 1968), 283.

32. Vary T. Coates and Bernard Finn, *A Retrospective Technology Assessment: Submarine Telegraphy; The Transatlantic Cable of 1866* (San Francisco: San Francisco Press, 1979), 171–172.

33. Kenneth Baglehole, *A Century of Service: A Brief History of Cable and Wireless Ltd., 1868–1968* (London, 1979), 15–17.

34. D'Humy, *A Brief Outline of the Technical Progress*, 41.

35. George P. Oslin, *The Story of Telecommunications* (Macon, Ga.: Mercer University Press, 1992), 342–343.

36. Clifford H. Cramer, "Submerged Repeaters for Long Submarine Telegraph Cables," *Electrical Engineering* 70 (1951): 1517–1523.

3

Bridging the Gap: The Cable and Its Challengers, 1918–1988

Jonathan Reed Winkler

When asked to explain how we once communicated electronically with one another internationally in the years before cheap long-distance telephony and e-mail, many people are at a loss for words. Unless they directly experienced the old methods of making a transatlantic radiotelephone call or sending a cablegram to someone a continent away, they might well have no idea at all. Within the past decades, the collective memory and direct knowledge of how those things once worked has gradually begun to fade. One of the most fascinating things about the forward march of technological progress is the rapidity with which we forget how we used to do things before the way we do them now.

This chapter will survey the history of cables between 1918, when their value and vulnerabilities had become manifest to the great powers, and 1988, when a new generation of fiber-optic cables rendered oceans an irrelevant barrier for data transmission. The sequential rise of telegraph cable, shortwave radio, telephone cable, satellite, and fiber-optic cable demonstrates the curious resilience of the cable as the senior instrument of international communications in the twentieth century. A full examination of the various technological, political, military, and economic pressures on international communications in this period, though sorely needed, is beyond the scope of this chapter. However, a brief overview will lay out several avenues for further inquiry about the persistence of cables in the mid-twentieth century.

The history of cables from World War I to 1988 can be roughly divided into four phases. From 1918 to 1926 cables remained the principal method of international telegraphic communications, despite the growing use of longwave radio transmitters for point-to-point communications in Europe, Africa, Asia, and the Americas. After 1926 the use of inexpensive vacuum tube transmitters for shortwave long-distance telegraphic communications severely cut into the market share and revenue of the major cable companies. They made some attempt to innovate and improve their operations to meet this challenge. It nearly proved to be too much. A wave of consolidations and mergers then reshaped the industry landscape. The Great Depression and war held further development in abeyance until 1956, when the first transatlantic telephone cable, TAT-1, became operational. The product of a joint venture by two previously minor participants in the cable field, AT&T and the British General Post Office, it was superior to the older cables by an order of magnitude. Successive designs of coaxial telephone cable provided ever more channels of communication. The advent of satellite-relayed telephone circuits beginning in 1965 posed the same threat to telephone cables as shortwave had done to telegraph cables a generation earlier. To keep pace with satellites, engineers designed subsequent cables with ever greater bandwidth. By 1988 the next class of cables had arrived to meet the satellite challenge, and fiber-optic cables now replaced the coaxial cables of the past thirty years.

The End of the Telegraph Cable Era, 1918–1956

A technology born of the 1850s, the submarine telegraph cable remained the dominant method of transmitting data electrically across vast ocean distances through the early 1940s. Steady advances in radiotelegraphy and telephony, particularly after the move into the shortwave portion of the spectrum, posed a continued threat to the cable in the early twentieth century. Though its proponents correctly argued that the radio could never replace the cable outright, the end of the telegraph cable era meant that the cable companies could no longer operate independently in the face of radio's challenge.

Built before World War I for commercial as well as strategic reasons, the global cable system just barely managed to serve the needs of the Allies during the war. The increased load of diplomatic, military, and commercial traffic, and the censorship over all private and commercial messages, had caused severe, widespread operational disruptions throughout the war. The information flow further constricted as occasional cable attacks or the deferred maintenance in submarine-threatened waters together degraded the cable network. Although the cabling public and concerned government officials wanted to ensure that an adequate cable system would be available after the war to assist the flow of commerce, little improvement seemed likely. The pressures of the war, the financial constraints on new investments, and the limited supplies of gutta-percha and new cable hampered further development by the cable companies. Within Britain the Eastern group of cable companies prepared to lay additional cables after the war rather than seek a technological leap in operational techniques or cable design to meet the anticipated jump in demand. In the United States the impetus for improvement came from a different quarter, which saw innovation as the only feasible solution to the cable problem in the absence of additional cable supplies.

During the war, allied military and naval officers had pressed for construction of radio stations to alleviate the overload on the cables. The U.S. Navy installed high-power radio stations to bridge the Atlantic and link the United States, Great Britain, France, and Italy while the cable companies repaired their damaged cables. In the Pacific, U.S. and Japanese commercial interests and the Japanese government were pressuring the United States to relax the wartime ban on transpacific commercial radio traffic to ease the incredible burden on the Pacific cables. Delays on these cables, when the lines were not broken, reached ten to fifteen days.[1] Having recognized the great importance of these cables, senior officials in the State, Navy, and War departments instructed the U.S. Commercial Pacific Cable Co. to lay a new line or improve the operation of the cable.

When the Commercial Pacific Co. curiously declined to improve its cable on its own, U.S. Army Signal Corps and State Department officials convened a committee of the country's best electrical engineers. This group was to consider the transpacific cable situation and devise the technical solutions that the cable company was unwilling to do. Led by Gano Dunn, head of the J. G. White Engineering Company and one of the founders of the National Research Council, its members included F. B. Jewett of General Electric, Arthur Kennelly of Harvard University, and J. J. Carty and Bancroft Gherardi of AT&T. None of the major cable companies was involved in the research, but each reluctantly provided proprietary information to the committee when asked.

Meeting from September to December 1918, the Dunn committee investigated how to increase the cable's operational speed. Dunn had the head engineer of Muirhead & Company, the leading cable equipment manufacturer in Britain, brought over to provide advice. After much research at the Western Electric laboratories, the committee advanced several important ideas. The fruit of this research was Dunn's recommendation that special signal magnifiers be installed at cable stations in San Francisco, Hawaii, Midway, and Guam. There were four such magnifiers in the United States at the time: two "Heurtley" examples held by the Alien Property Custodian and two "Selenium" at Western Union laboratories.[2] Installation of these on the Pacific route would boost the speed of that cable by 30 percent. If the government could arrange their installation on faster cables in the Atlantic or to South America, even higher results were possible. Dunn also urged that Bruce relay repeaters be installed at the island stations to upgrade the operation still further.

As the federal government had nationalized the cables during the final months of the war, the improvements could easily be made despite the reluctance of Commercial Pacific.[3] State and War Department officials pushed for continued research into the technical problems of cables and signaling after the war. Private groups, including the Council on Foreign Relations as well as shipping firms and international banks, also pressed for improved cable access for the United States. A chief objective was the development of an artificial substitute for gutta-percha latex. Supplies were inadequate, and current substitutes—like rubber—had various limitations. Despite support from senior government officials for continued research, insufficient funding and the incapacitation of President Wilson prevented any significant research or breakthroughs.[4] Therefore, the initiative for further cable development passed back to private companies.

Improvements in cable technology to this point had largely come from British firms—Muirhead & Company and others—that were the leading designers and manufacturers of cable equipment. After the war, Western Union took steps to build up its own research and development laboratories for the first time. However, as Bernard Finn shows in this volume, the cable companies remained largely conservative in their laboratory activity. In 1920 Western Union moved the ocean cable headquarters back from London to New York. Its Engineering Department took over research and design work previously done by the leading British firm Clark, Forde & Taylor. Spurred perhaps in part by the Dunn committee, Western Union also made improvements to operations and installed Heurtley magnifiers and the cable rotary repeater on its main cables by 1920. This equipment significantly improved cable operation, raising the speed to 100 words per minute and reducing the need for manual retransmission.[5]

The apex of submarine telegraph cable design occurred in the 1920s. In the United States, the cable companies were able to keep pace with the advancements in longwave radiotelegraphy. As a result, they were in a stronger position to cope with the subsequent rise of commercial shortwave radio as well. The first major advancement was in the design of the cables themselves. Bell Laboratories at AT&T had produced a nickel-iron alloy called Permalloy for use on telephone repeaters, and in 1923, together with Western Union, they developed a Permalloy-wrapped cable capable of carrying signals three times faster than existing cables. In these "loaded" cables, the Permalloy acted as a magnetic inductor, helping offset the deleterious effects of the electrical capacitance. Oliver Heaviside had shown this effect to be theoretically possible in the late 1880s, but it had required significant advances in metallurgy

Table 3.1
Transatlantic Telegraph Cables

Laid	Cable Route	Owner/Operator	Closed	Notes
1873	Newfoundland-Ireland	Anglo-American	1963	Control to Western Union 1912
1874	Newfoundland-Ireland	Anglo-American	1963	Control to Western Union 1912
1875	Nova Scotia–Ireland	Direct US (Western Union)	1962	To Great Britain 1920
1879	Massachusetts–Brest, France	CFCT	1929	
1880	Newfoundland-Ireland	Anglo-American	1949	Control to Western Union 1912
1881	Canso, Nova Scotia–Cornwall	American Telegraph and Cable (Western Union)	1963	
1882	Canso-Cornwall	American Telegraph and Cable (Western Union)	1963	
1884	Canso-Ireland	Commercial Cable	1963	Diverted to St. Johns 1909–1910
1894	Canso-Ireland	Commercial Cable	1963	Diverted to St. Johns 1926
1894	Newfoundland-Ireland	Anglo-American	1963	Control to Western Union 1912
1898	New York–Brest, France	CFCT	1962	
1900	New York–Azores–Germany	DAT	1962	Cut by Allies 1914
1901	Canso-Azores-Ireland	Commercial Cable	1962	
1904	New York–Azores–Germany	DAT	1962	Cut by Allies 1914
1907	Canso-Ireland	Commercial Cable	1962	Diverted to St. Johns 1926
1910	New York–Newfoundland–UK	Anglo-American	1963	Control to Western Union 1912
1923	Canso-Azores-UK	Commercial Cable	1962	
1924	New York–Azores	Western Union	1966	Linked to Italcable
1926	Newfoundland-Cornwall	Western Union	1966	
1928	Newfoundland-Azores	Western Union	1966	

CFCT = Compagnie Française des Câbles Télégraphiques
DAT = Deutsche-Atlantische Telegraphengesellschaft

and electrical theory before the cable companies could put it into practice. Commercial Cable and its British design engineers were skeptical of the idea. Instead, they went ahead with plans for a huge cable (1,100 pounds of copper per nautical mile) to the Azores in 1923. Western Union, more patiently, tested the Permalloy design and spent much of its cash reserves to lay three new loaded cables in 1924, 1926, and 1928 (see table 3.1). With speeds of up to 300 words per minute (wpm) on these new cables, the increased signal capacity still barely met the growing demand for international telegraphy. Western Union found it had to continue to keep the older cables in operation, despite the growing challenge from point-to-point radio. As the shortwave operations cut deeper into the international telegraphy market, the cable companies again responded with new signal shaping, vacuum tube amplification, and time division multiplexing equipment that permitted operational speed of up to 480 wpm on the newest cables.[6]

The move into the shortwave bands and use of vacuum tube transmitters in the late 1920s finally enabled long distance radio communications to threaten the predominance of cables. Between 1924 and 1926 the Marconi's Wireless Telegraph Company of Britain and the Radio Corporation of America (RCA) adopted transoceanic shortwave radiotelegraphy. AT&T introduced commercial intercontinental radiotelephony in 1927. The relatively low cost for shortwave transmitters, receivers, and other equipment allowed the radio companies to offer their services for a far lower rate than the cable companies could afford. For political as well as commercial reasons this spawned a tide of mergers in Europe and the Americas that by 1929 consolidated the industry into a small handful of players. Economic depression and war spared further depredations to the cables, but the rise of shortwave changed forever the companies that operated them.

The shortwave end of the spectrum had initially been thought of as useful only for very short-range transmissions. Research by Guglielmo Marconi and the U.S. Navy demonstrated in 1924–1925 that the signal bounced off the ionosphere, allowing transmission over very great distances. With improved transmitting and receiving equipment afforded by the vacuum tube and other developments, shortwave was not only commercially feasible but also far easier to use than existing longwave systems.[7] Marconi's company announced in May 1924 its success in sending and receiving shortwave signals between Cornwall and Marconi's yacht in the South Atlantic and elsewhere over previously unbelievable distances. Other firms and government laboratories undertook similar research. By 1926–1927 the Marconi Company had implemented its plan for using shortwave on the proposed British government's Imperial Wireless Chain. In November 1926 the United

Kingdom–Canada circuit opened. By June 1928 the United Kingdom and Australia were in direct, reliable radio connection for the first time. After initial skepticism about the efficacy of the undertaking, RCA announced in October 1926 that it too would open radio circuits on the shortwave end of the spectrum. Its new station in the Philippines opened in June 1927, with shortwave equipment rather than the more expensive longwave transmitters previously planned. The French opened similar stations in Southeast Asia.

Shortwave transmitters were appealing due to their low cost and small power requirements relative to the distances signals could travel reliably. The new vacuum-tube-based transmitters, operating on low-kilowatt power, cost only one-twentieth the investment of the latest longwave transmitters. This made radiotelegraphy extremely cost effective against the cables. The U.S. Navy estimated for Congress in 1928 that a typical transatlantic shortwave radio service built by RCA would require, at a minimum, two stations, at cost of $700,000, with annual maintenance and personnel costs at $250,000. A new cable on the same route ran approximately $8,000,000, with annual maintenance costs of nearly $1 million. Because the companies were investing for the long term, such as twenty years, the total costs worked out roughly to $34,380,000 for the cable and $5,700,000 for the radio, regardless of the traffic load handled. A modern radio station providing shortwave radiotelegraphy would thus cost only one-sixth as much as a cable as an investment over the short or long haul.[8]

Further developments in international voice communications at this time also complicated the situation for the cables. After much development work from 1915 to 1926, AT&T opened the first commercial radiotelephony circuits by shortwave to London in 1927. Connections to Europe came the following year. Although expensive to use, they allowed for more rapid communication between high-paying customers than the fastest urgent-rate telegraphic messages. Subsequent innovations in transmission, antennae, and operation allowed AT&T to make 15 to 16 channels available for conversations. Also in that year Juan Trippe's nascent Pan American Airways, ferrying inexpensive airmail to South America and across the Pacific, threatened to whittle away at the slower-rate messages as well. The cable companies thus faced significant challenges to their position from two sides.

Despite the improvements to cable operations and the introduction of the loaded cable, shortwave radio continued to cut inexorably into the market the cable companies had held firmly for more than a half century. Radio competed with the existing cable networks on the profitable routes. They could also exploit the niche market for connections to less lucrative markets, such as landlocked countries that the cable companies could not reach. By the end of the decade RCA had gained 30 percent of the transatlantic market and fully 50 percent of the transpacific market for international telegraphic messages. This proved to be so contentious for Western Union that by 1929 it reluctantly lowered its transatlantic cable rates for the first time since 1914 just to keep pace.

In the British Empire, the Marconi-installed Imperial Wireless Chain quickly diverted up to 30 percent of the international traffic between the United Kingdom and Australia. Traffic managers suspected that the numbers were probably higher. In some instances, such as deferred press traffic, they estimated that it reached as high at 50 percent. Indeed, it was apparent to some informed commentators within months of the opening of the service in 1926 that the Eastern group cables would not be able to keep up on the major transoceanic routes without shedding the weaker cables or fighting a ruinous rate war. The crowning realization that radio had supplanted cables came in November 1929, to RCA's pleasure and the dismay of the cable companies. A powerful seaquake in the North Atlantic severed 12 of the 21 cables in short order. Cable officials had to transfer their traffic load over to the shortwave circuits until the cables were restored. The shortwave network was robust enough to handle the additional traffic with relatively minimal disruptions in service.

The leading cable powers reacted differently to the political and economic difficulties raised by the technical strengths of radio over cable. In Britain, John Pender's Eastern group could not match the government-backed Imperial Wireless Chain on the more remunerative long-distance routes. As early as the fall of 1927 Pender began to hint to the shareholders that the firm might voluntarily liquidate itself rather than suffer decline and collapse. This prospect horrified government officials, who were loath to lose the strategically important cables. Discussions among the various Dominions and colonies prompted the Imperial Wireless and Cable Conference of January 1928 to consider the overall threat to the imperial and commercial cables from the growing shortwave service. The Conference recommended approving a Marconi-Eastern merger as well as compelling the new company to acquire both existing government cables and the beam service. The result would be a massive, government-overseen communications monopoly with a truly global reach. Marconi and the Eastern group then announced in March 1928 that they had under consideration a plan for merging. This proposal, prompted in part by hints of cable and radio mergers in the United States, was far more acceptable to the government. By April 1929 the new Cable and Wireless Ltd. holding company came into existence. The formation of this new telecommunications company from the old cable and radio companies finally separated the method of delivery from the message itself. The merger brought an operational change that saved cables, when technological innovation could not, by preserving them in place as part of a larger network rather than forcing them to compete separately against the piecemeal attacks of radio.

As in the United Kingdom, the principal radio company in the United States put pressure on the cable operators, but here the threat from RCA was less immediate. The existing cables were all reasonably profitable (although the Pacific cable was notoriously slow). Unsupported by subsidies to meet national defense needs, they were therefore less vulnerable to rate wars. Unlike Pender's Eastern group, Commercial Cable and Western Union cable lines did not have to share profits with a government-run domestic telegraph network for distributing cables within the United States. In the late 1920s some consolidation occurred in the United States but it followed a different pattern. What did, and did not, occur did so for the same convoluted political and economic reasons that had compelled the formation of Cable and Wireless.

The first phase of consolidation was not directly related to the radio threat. Sosthenes and Hernand Behn's International Telephone and Telegraph Company (ITT) had grown from telephone operations in Cuba and Puerto Rico to one of the largest international telephone holding companies in the world by 1926. Behn had then turned his attention to cable properties for additional growth. After much negotiation Behn acquired All America Cables, the most profitable of the firms, in April 1927 to complement ITT's larger South American holdings. On July 1, 1928, ITT then added the even larger corporate shell of the Mackay Companies. This included not only the Commercial Cable Company and Commercial Pacific but also the new Mackay radio subsidiary formed of Federal Telegraph, RCA's only major competitor. The component companies still worked as independent units, but by the end of 1928 ITT had become a major force in international communications.

The second phase of consolidation took place during the unsuccessful merger talks of 1928–1929. These discussions revealed how concerned the cable companies had become about the threat of radio. The successive announcements by Behn's ITT in 1927–1928 and the threat of a powerful American combination had played no small role in driving the British cable and wireless alliance. Now it was the threat that the conjoined Marconi and Eastern companies might overpower the American cable companies that overshadowed everything. Western Union had spent its large cash reserves on the new loaded cables of 1926 and 1928. In the face of the British combination and RCA's increased share of the transatlantic service, this investment was looking increasingly ill-timed. As profits began to slump in the domestic market as well, Western Union officials looked to shedding their international cable interests as one way out. Some form of combination with RCA was the most opportune combination, and Carlton found willing ears there. Although the U.S. Navy had pushed the formation of RCA by General Electric initially for international communications, the domestic market for radios had taken off so strongly that by 1928 only 7 percent of RCA's revenue came from its long-distance overseas operations.

In February 1928 Newcomb Carlton and RCA's Owen Young sailed together to Britain to investigate the merger talks there. On the lengthy voyage they also discussed their own consolidation possibilities. Western Union controlled 50 percent of the international cable business in the United States (and ITT the other half). A combination with RCA's international division would have created a powerful U.S. cable and wireless firm. The negotiations went so far that their respective bankers—Kuhn, Loeb for Western Union and Morgan & Co. for RCA—negotiated financial terms through the rest of 1928. General Electric, the parent firm of RCA, acquired a great deal of Western Union stock, while RCA devolved its international business into a smaller unit, RCA Communications, Inc., in preparation. At the last minute, to Young's disappointment, Carlton backed out of the proposal. The reason is still not clear today, but legal obstacles likely played a part. Certain provisions of the Radio Act of 1927(also known as the White Act) restricted cable and radio companies from becoming involved in each other's activities, specifically to prevent a unification of cable and radio interests under hostile control. It is possible that Carlton believed that there might be legislative changes that would allow Western Union and RCA International to merge. When these did not appear, Carlton then backed out rather than risk congressional opprobrium.[9]

Disappointed, Young and RCA then turned to Behn at ITT, who was very interested in acquiring RCA's strong international radio operations to add to ITT's portfolio. On March 30, 1929, RCA and ITT jointly announced that they had formed a tentative plan to merge, should it become legally possible. The spate of government-encouraged amalgamations in Britain, France, Germany, and Italy and the competition they foreshadowed helped to drive this urge to consolidate. However, in October the stock market crashed. ITT's stock, which was to have been a core part of the merger, took a substantial drop. In December, Congress considered the larger question of monopolies and strategic needs in international communications but took no steps to unravel the knots in the law that prevented the consolidation of the cable and wireless industries. Although Behn and Young periodically discussed the deal further, there the matter remained as the Depression expanded.

It was the Depression and war that, as damage from the shortwave grew, granted the cable companies reprieve in the United States. Initially cable traffic continued to rise, with ITT seeing a 15 percent growth in telegraphic cable traffic in 1930, although many business users saved money by forgoing "urgent" for "regular" or "deferred" rates. This changed after 1931, when the decline in traffic and the difficulties of the financial market led ITT and RCA to call off their proposed merger. By 1933 the situation looked bleak. The American companies began discussions with Cable and Wireless over new traffic-sharing arrangements and other possible reductions in expenses in order to remain solvent. Between 1930 and 1940, the cable companies fell from handling 82 percent of the international telegraphic traffic to handling 57.3 percent, while RCA, the dominant radio firm, saw its share rise from 18 percent to 42.7 percent.[10] Business users simply were choosing the most direct and most inexpensive method to send their messages. Demand surged again only in the final crisis years of the 1930s. With the fall of France in 1940 and the loss of European traffic, Commercial Cable and Western Union suffered a brief steep decline. Commercial Pacific and All America, meanwhile, both experienced a steady rise in traffic over their lines at least through 1941. When war finally came, the United States and Great Britain, controlling the vast majority of the world's cables, reintegrated the commercial networks available back into the military communications systems, as had been done in World War I. While radio proved a crucial technological innovation for the operational and tactical command of forces in the field, the cables again offered secure strategic intercommunication around the globe. Radio signals could be and were monitored on both sides; telegraphic cable messages were far more difficult to intercept.

Throughout the war, senior government officials in the United States and the United Kingdom continued to contemplate how to protect the cables against the threat from radio. Cable and Wireless would remain a government-sanctioned monopoly, but it would eventually be nationalized and brought closer to the government. In 1943 Western Union absorbed the domestic operations of its now-bankrupt rival, Postal Telegraph. Under Federal Communications Commission rules, Western Union now had to divest itself of the North Atlantic cables it gained from Postal. Given the circumstances, such a move was neither practical nor wise, and the FCC issued a unique waiver instead to postpone the issue until after the war. By this time two different ideas of a government-supported consolidation emerged. One, oriented around the traffic carried, was to create two separate companies: telephony under one company, telegraphy under another. The other, framed around the method used, would provide for three separate companies, one each for radio telegraphy, telephony, and cables. By 1944 the U.S. Navy pushed for a single unified company to handle cables and radio together, but the matter remained unresolved as the war came to an end. In the absence of any technical innovation to raise the capacity of cables to the level of radio while reducing the investment costs, radio continued to make inroads on the cable markets.

The Copper Telephone Cable Era, 1956–1988

Twenty-seven years separated Western Union's laying of the last major telegraph cable and the laying of the first transatlantic telephone cable (TAT-1) in 1955. The disruption wrought by the Depression and the war forestalled any major new commercial cable development, although much engineering work done for the war affected the later development of the new cables. Significantly, further innovation would come not from the telegraph companies but from those working on telephony. In fact, new technology came almost entirely from the telephone giant AT&T and its Bell Laboratories. Unlike previous cables development, successive advancement in cable design would bring cables prodigiously more powerful than anything contemplated before 1955.

Telephone cables had been in use to cross rivers and relatively narrow bodies of water such as the English Channel since the late nineteenth century. So long as the underwater distance was short, the degradation of the signal during its passage through the cable was not sufficient to render the conversation inaudible. Over longer distances, however, the signal steadily diminished until it was useless. A telephone signal uses a broader band of frequencies and greater bandwidth than does an ordinary telegraph signal. Since the effect of electrical capacitance (a consequence of the surrounding sea water) is a function of frequency, the different parts of a telephone signal are selectively attenuated. Over a significant distance the signal becomes weaker to the point of incomprehensibility. Loading, as mentioned above, provided some relief. But the real solution, readily recognized, was to amplify the signal at intervals along the length of the cable.[11] Bell Laboratories engineers worked aggressively on the amplification problems throughout the 1920s, laying the groundwork for the major advancements of the 1950s. The Army Signal Corps particularly encouraged efforts to improve submarine cables and to construct a cable manufacturing plant as a means of severing dependence on foreign suppliers.[12]

The continued use of AT&T's expensive radiotelephone system launched in 1927 demonstrated that the demand existed for a profitable transatlantic voice system, if the technical challenges could be worked out.[13] AT&T representatives proposed to the General Post Office (GPO), which held the monopoly on international telephony in Britain, that a continuously loaded cable might be used for an unamplified, single-voice channel cable between North America and Britain. By 1929 AT&T officials were sufficiently confident in the likelihood of a viable telephone cable for long-distance use that on November 4 company agents obtained a concession from the Newfoundland parliament to land a telephone cable connecting with Europe. Even Newcomb Carlton of Western Union, whose telegraph cables AT&T might endanger, willingly conceded the advancement as a valuable contribution and important counter to the power of the newly merged British cable and radio interests.[14]

The stock market crash, the financial depredations of the Depression, and war forestalled AT&T's ambitious plans but did not end them completely. As the project simmered from 1929 through the end of the war, successive advances in telephone equipment and technology continued to alter the design of the eventual cable. Bell Laboratories' engineers had decided that repeater technology, with component vacuum tubes and other electronics, would be necessary to boost the signal along the way. These repeaters would need to be reliable enough to have a projected service life of twenty years underwater without the need for regular maintenance or replacement. In addition, the methods for working multiple phone calls on land—known as multiplexing—might be used as well on the underwater cables to scale up the volume of traffic, but only with additional technical developments.

The development of smaller, more reliable vacuum tubes for military use during World War II was the breakthrough that at last would allow repeaters and transatlantic telephone cables. The GPO first tested vacuum tubes as repeaters on telephone cables spanning the Irish Sea in 1943. By this point the plans at AT&T had been worked out for a 12-channel cable, using repeaters at intervals of 50 miles. This would require about forty repeaters along the transatlantic portion. Bulges in the cable due to these repeaters and their housings would affect the way in which the cable might be laid. As AT&T wanted to lay the cable in one single operation for safety reasons, this meant that the repeater should be not much larger than the cable and capable of being wrapped up in the storage drum along with the cable. Any other way, in the eyes of the cable engineers, would have required new ships or new operating techniques that were beyond the experience of the crewmen and conventional methods for laying cable. The result was an initial design of an 8-foot-long repeater housed in a 70-foot-long component section. These size restrictions constricted the equipment, limiting the signal bandwidth, preventing the cable from operating in duplex, and necessitating that a second cable be laid to carry the traffic in the opposite direction. Given that this was completely new cable technology, it was a costly solution—but the best given the limitations of the repeater. AT&T put this method to the test in the 1950 Key West–Havana cables.[15] The two simplex cables, each 120 nautical miles long, had repeaters every 40 nautical miles to boost the signal for the 24 channels. With the technology fully viable, attention turned to the North Atlantic.

In 1952, negotiations resumed between AT&T and the GPO for the transatlantic telephone cable. Four years later the first transatlantic telephone (TAT) cable was operational. Like AT&T, the GPO had also gained experience during the war with vacuum tube repeater technology and by now had laid several repeatered cables across the Irish Sea and to the Continent. GPO repeaters were stiff, not articulated, and this different method required that the ship laying cable stop at intervals in order for the crew to attach the repeater to the cable section drawn out of the tank. After negotiations about which method would prevail, AT&T and the GPO signed an agreement on November 27, 1953, to lay the cable and to use the Bell repeater system for the transatlantic portion. As originally envisioned, the cable would carry 36 channels (already a threefold increase over the plans of the previous decade), and AT&T and the GPO would lay it jointly. When finally opened for commercial use on September 25, 1956, the cables made use of 51 vacuum tube repeaters with about one every 30 miles. This initially gave 36 4-kHz channels, later modified to 48 3-kHz channels and then to a maximum of 51 channels for voice communications. Compared to the 15 channels in the shortwave-based radiotelephony, it was a fantastic step forward.[16]

The capacity of the initial TAT-1 telephone cable compared to the existing cables was astounding. Its arrival was the death knell of the telegraph cable system. TAT-1's 51 telephone channels provided the equivalent of 459 telegraph channels (a 1:9 ratio), vastly more than all the existing North Atlantic cables together. At 50 wpm on each channel, or in total 22,950 wpm telegraphically, TAT-1 collectively supplied 48 times the best message rate of the single fastest prewar cable (480 wpm), or roughly six times the total capacity of all the prewar cables (3,853 wpm).[17]

The old telegraph cables were still usable, however, and they remained in service for several more years. The cable companies decommissioned all of the pre–World War I cables by 1962–1963. The four transatlantic lines laid in the mid-1920s came out of service in 1966.[18] The major cable companies, which had been a fixture in the lives of coastal Nova Scotians and Newfoundlanders, left soon after. In 1955 Western Union pulled out of Canso, Nova Scotia, where it had been since the 1870s. Commercial Cable shut down its operations there as well in 1962, after almost eighty years. The future belonged, or so it seemed, to the coaxial telephone cable.

Successive coaxial telephone cable design, incorporating the continuing advances in this new technology, produced three more gigantic leaps in capacity over the following years. Research on the next generation of cables began even before the TAT-1 became operational. TAT-2, laid in 1959, was similar to TAT-1, with two cables providing unidirectional traffic. Between 1959 (TAT-2) and 1963 (TAT-3), the design engineers had resolved the problems of size and cable laying necessary for putting down a single cable with two-way repeaters. From this came CANTAT-1, linking Scotland to Newfoundland in 1961, which was the first single-cable system, boasting 60 channels. In order to increase the available bandwidth, engineers then turned to reducing the spacing for channels from 4 kHz to 3 kHz. By squeezing tighter the range within which the voice in a telephone call operated, they could replace 12 channels with 16 with a minimum of disruption to the conversation. Engineers also devised time assignment speech interpolation (TASI), a technique adapted from landlines where improved switching and detectors could make use of the dead air in conversations. This allowed for more efficient use of the channels. With increasingly powerful repeaters and improvements to the entire cable operation, distance ceased to affect where the cable might be laid. In the Pacific, a new telephone cable replaced the venerable Imperial telegraph cable from Canada to Australia in 1962–1963. Here, the longest single section was 3,073 nautical miles, between Fiji and Oahu, a fantastic distance that necessitated 118 repeaters but did not prevent the cable from having 84 channels. Commercial desires, rather than strategic or technical requirements, now dictated where a cable might land.

The next generation of cable design, incorporating greater bandwidth within the cable and thereby providing still more channels, took ten years to develop. The result was TAT-3, laid in 1963. It boasted over 128 channels. With vacuum tube repeaters every 20 miles, cost and not signal degradation now affected the length of a cable. TAT-3 went directly from Tuckerton, New Jersey, to Great Britain. The old landings in Canada and Ireland were irrelevant. A second cable of this type, TAT-4, linked France to New Jersey in 1965. These four TAT cables together provided 375 channels, but one of the new cables alone had more capacity than TAT-1 and TAT-2 together. The interval had also given AT&T the chance to design a new cable ship. Unlike earlier cable ships, the CS *Long Lines*, launched in 1961 and used to lay TAT-3, could handle the bulkier cable and the special problem of passing repeaters over the bow.

Improved and relatively inexpensive access to transatlantic telephony sparked a boom in demand, which in turn raised pressure for still more cables. To this point the repeaters had been vacuum-tube-based, even though on land the transistor had become a critical component in new electronic devices. Engineers had been reluctant to commit to the transistor for submerged cable repeaters because early transistors had not been robust enough to meet the twenty-year reliability requirement already imposed on the vacuum tubes. By the late 1960s, however, transistors had reached the stability that cable engineers sought. After tests had shown them to be successful on short cables, transistors became a part of the plans for TAT-5. The result was another tremendous increase in capacity. This cable, laid in 1970 between Spain and Rhode Island, had a bandwidth of 6 MHz, with repeaters every 10 miles. The designers had anticipated 720 channels but, to their surprise, achieved 845, more than twice the existing transatlantic capacity combined. The CANTAT-2 cable, laid in 1974, boasted another 1,840 channels.

The final advancement in coaxial telephone cables and the peak of the technology came in the mid-1970s (see table 3.2). Satellite communications had begun to make available circuits for international telephone traffic from

Table 3.2
Transatlantic Telephone Cables

Laid	Name	Cable Route	Channels
1955–1956	TAT-1	Scotland-Newfoundland	51
1959	TAT-2	France-Newfoundland	48
1961	ScotIce	Scotland–Faero Islands–Iceland	26/24
1961	CANTAT	Scotland-Quebec	80
1962	IceCan	Iceland-Greenland-Newfoundland	24
1963	TAT-3	Great Britain–New Jersey	138
1965	TAT-4	France–New Jersey	138
1970	TAT-5	Spain–Rhode Island	845
1973	BRACAN	Brazil–Canary Islands	160
1974	CANTAT-2	UK–Halifax, Nova Scotia	1840
1976	TAT-6	Rhode Island–France	4190
1977	Columbus	Venezuela–Canary Islands	1240
1982	Atlantis	Portugal-Senegal-Brazil	1380
1983	TAT-7	UK–New Jersey	4200

1967 onward, and this affected new cable development. The monster cables envisioned to compete with the satellites would have a bandwidth of 29.5 MHz and would provide 4,000 channels by using transistorized repeaters every 5.1 nautical miles. Engineers were now able to condense five conversations into one voice circuit, but in order to do this they had to increase the repeaters to stem the loss on the higher frequencies. Repeaters were expensive and in turn raised the cost of the cable significantly. TAT-6 cost $179 million, but the increased number of channels brought the per-channel cost down to $45,000, less than half that of TAT-5. Another cable of this design, TAT-7, followed in 1983. AT&T designers planned for a follow-up cable, handling 16,000 channels, but it would have required a thousand repeaters.

As the bandwidth increased to accommodate the higher frequencies needed on the added channels, the cable also became physically thicker. TAT-6, for instance, was more than two inches across. Engineers began to worry that the cables could become too big and heavy for one cable ship to handle. The arrival of fiber optics, however, pushed research into another direction. AT&T designers shelved plans for any further monster coaxial cables.

From 1956 to 1965 telephone cables offset radiotelephony for connection on the most important routes across the Atlantic and the Pacific. The great expense relative to the traffic carried, however, slowed their introduction to other regions of the world. This situation changed again with the introduction of radio signals transmission relayed off of satellites. Contemplated since 1945, feasible from 1957, and tested from 1962 on, satellites for communications became a reality in April 1965. At that point the federally chartered Communications Satellite Corporation (COMSAT) successfully deployed Early Bird, which provided 150 "half-channels" (or roughly 300 conversations) for communication. This was still more powerful than the latest transatlantic cable at the time, TAT-4, which had 138 two-way channels. Other countries began to build ground receiving stations to link their national communications grids with the international satellite transmissions. In 1964 COMSAT, together with its counterparts in other countries, formed the collective International Telecommunications Satellite Organization (INTELSAT). Over the next decade, this organization provided several series of satellites for international communications in different regions of the world, often to places where the new telephone cables had not yet been laid. The INTELSAT-III series had by 1969 provided global coverage for telephone and data communications—"wiring" the world with satellites—within a remarkable four years.

Early Bird and subsequent satellites provided a significantly greater number of channels at much cheaper rates than the transatlantic telephone cables. This disparity continued from 1965 until 1988 and the first fiber-optic cable. Yet satellite communications were never so perfect that they pushed aside the telephone cables completely. The high orbit of the satellites imposed a transmission delay of up to a quarter of a second each way. Satellites were also vulnerable to signal disruption from exceptional solar activity—a static problem that had also troubled the earlier radiotelephony. Poor connection problems, resulting in feedback or echoed signals, could plague the satellite call as well.[19] Expensive to build, impossible to maintain, and ultimately disposable, satellites also had a lifespan that, although long, limited their utility compared to the cables. Still, the larger number of channels available through the satellite relative to the large investment costs made it worthwhile for countries in many regions to connect to their neighbors by satellite rather than by cable. Thus the 1970s saw the proliferation of both cables and satellites. As with radio and cables by the 1940s, the method of transmission had become less important than the message carried. Cables were not doomed to being replaced; rather, they would again work in concert with radio.

The introduction of TAT-8 in 1988 and the expansion of global fiber-optic cables since then has changed the international communications situation dramatically. More powerful than both the older coaxial telephone cable lines and the satellite-relayed radio communication, these new cables have again placed cables "on top" over radio.

The recurring pattern evident through the twentieth century—first of telegraph cables, then radio, coaxial cables, satellites, and finally fiber optics—raises the question of why cables have been so resilient. Cables survived because they successively offered greater capacity despite their immobility and the tremendous installation costs. Greater capacity spurred greater demand, and this in turn ensured continued, if declining, profitability for the older systems in the face of newer advancements. This cycle allowed the telegraphic cables to survive the initial radio challenge. The spate of development from TAT-1 in 1956 to TAT-8 in 1988, and the continual growth in international communications that accompanied these, were too quick for the satellites to overwhelm the cables. Cables and radio, in the end, have complemented one another.

Cables also survived because it was in the interest of the leading powers in communications, first Great Britain and then the United States, to ensure that the companies operating the different systems did not ruin themselves competing against one another. The cables continued to have imperial and strategic significance that was disproportionate to their age or capacity. In Britain, protection of the cables necessitated the unification of the cable and radio firms into a conglomerate, Cable and Wireless. In the United States, preservation of the communications networks meant the tacit sanction of near-monopolies by the major cable and radio firms to preserve the redundancy of having more than one method of communicating with the world. The military and governmental needs for the secrecy inherent in cable-based communications ensures that the strategic importance of global cable networks first envisioned by Great Britain before 1914 continues to be of chief importance for the United States into the next century.

Notes

1. At this point it would become cheaper to rely on surface mail, as the transpacific steamers from San Francisco to Japan took about 15 days to make their journey.

2. The provenance of the two magnifiers held by the Alien Property Custodian is not known, although it is reasonable to presume that the German Atlantic Cable Company at New York owned them.

3. The Commercial Pacific firm may have been reluctant to improve the quality of its operations because a British and a Danish cable firm held large shares of ownership in the company. These two companies already controlled cable communications in the Far East and stood to lose from any improvement in U.S.-Asian communications.

4. Interim Report of Interdepartmental Committee on Communications, January 25, 1919, "International Communications Conference" folder, box 30, Norman Davis Papers, Manuscript Division, Library of Congress; Breckinridge Long to Julius Lay, February 26, 1919, file 811.73/151, State Department Decimal Files, Record Group 59, National Archives and Records Administration, College Park, Maryland.

5. "Submarine Telegraphy," in *A Brief Outline of Technological Progress by Western Union, 1910–1934*, vol. 2, box 2, Western Union Collection, Archives Center, National Museum of American History, Smithsonian Institution, Washington, D.C.

6. Vary T. Coates and Bernard Finn, *A Retrospective Technology Assessment: Submarine Telegraphy; The Transatlantic Cable of 1866* (San Francisco: San Francisco Press, 1979), 156–162; "Submarine Telegraphy," 124–128.

7. Until approximately 1924 the theories about radio signal propagation held that the higher longwaves (or lower kilocycles) were the most effective bands available for communicating over distances greater than 1,500 miles. This in turn led to strategic worries that a country needed to obtain exclusive use of enough longwaves to communicate independently with other powers and overseas possessions.

8. Testimony of Captain S. C. Hooper, USN, in Senate Committee on Interstate Commerce, *A Bill to Provide for the Regulation of the Transmission of Intelligence by Wire or Wireless: Hearings on S. 6*, 71st Cong., 1st sess., May 8 to June 7, 1929 (Washington, D.C.: Government Printing Office, 1930), 1:326.

9. Josephine Young Case and Everett Needham Case, *Owen D. Young and American Enterprise: A Biography* (Boston: David R. Godine, 1982), 461. Compare Robert Sobel,

ITT: The Management of Opportunity (New York: Times Books, 1982), 62–63. Sobel suggests that the banker Thomas Lamont convinced Young to reject Carlton's offer, while the Cases (Everett Case was Young's assistant) suggest that Carlton broke off the talks with Young and that Lamont expressed sympathy with Young's disappointment. A proper recounting of this tale, making use of governmental and military records, would be useful. Commander Stanford C. Hooper, founder of the U.S. Navy's modern radio program and later advisor to the early Federal Radio Commission, recounts in his personal memoirs that he acted several times to stop mergers within the cable and radio industry from behind the scenes, but this has not been brought forward properly in secondary accounts. See Hooper notes, "Biographical Material" folder, box 44, Hooper Papers, Manuscript Division, Library of Congress.

10. Figure from "Recent Activities of the International System, June 1933," box 111, Clark Radio Collection, Archives Center, National Museum of American History. See also charts in S.Rep. No. 769 (Study of the Telegraph Industry), 77th Cong., 1st sess. (Washington, DC: Government Printing Office, 1941).

11. The difference between the bandwidth necessary for telegraph messages and voice traffic can be demonstrated easily today. Transfer of files containing music over the Internet takes up greater time and capacity than that of a large e-mail with simply a text attachment.

12. Sosthenes Behn and others had laid a telephone cable between Key West and Havana in 1920, but it was not as advanced as the 1923 cable. Behn's company was the core of the eventual giant ITT.

13. Five minutes of airtime cost $75, a substantial cost that limited use to only the most pressing international business communications of wealthy clients.

14. Notes of meeting between Western Electric engineers and U.S. Signal Corps officers, March 16, 1923, "Western Electric" folder, file 095, Commercial Firms, General Correspondence of the Chief Signal Officer, Record Group 111, National Archives and Records Administration.

15. The Caribbean and its islands were often a test bed for new cable design. In addition to the Key West–Havana cable before TAT-1, AT&T laid new cables to Jamaica before TAT-3 and to the Virgin Islands before TAT-5.

16. Much of this section follows the work of R. D. Ehrbar, "Undersea Cables for Telephony," in Peter K. Runge and Patrick R. Trischitta, eds., *Undersea Lightwave Communications* (New York: IEEE Press, 1986).

17. On the capacity of telegraph cables, see Coates and Finn, *A Retrospective Technology Assessment*, 156–162.

18. The 1879 French cable, long inadequate for the task, had been mercifully shut down in 1929, and the 1880 Anglo-American cable stopped working in 1949.

19. Jeff Hecht, *City of Light: The Story of Fiber Optics* (New York: Oxford University Press, 1999), 204.

4

Fiber-Optic Submarine Cables: Covering the Ocean Floor with Glass

Jeff Hecht

Charles Kao proposed the concept of fiber-optic telecommunications in 1966, while working at Standard Telecommunication Laboratories, a subsidiary of the then-giant International Telephone and Telegraph Corporation (ITT). His initial vision was bold for the time but quite modest in retrospect. He thought that glass could be purified enough to transmit signals for about 10 kilometers, and made into fibers that could carry a billion bits per second (a gigabit) over that distance. His vision was to carry signals between local telephone switching offices in adjacent towns, a small niche where new technology was needed.

ITT's resident visionary, Alec Reeves, could see farther. Then in his mid-sixties, Reeves had invented digital pulse-code modulation, now used throughout the telephone network, in 1937. In a 1969 talk, even before the first low-loss optical fibers had been made, he thought about a fiber-optic submarine cable:

> In the much more difficult case of the sub-ocean routes, it may take up to about 20 years to produce repeaters with long enough average lives to give complete cables 5000 km long that will operate without maintenance for an average period of 20 years—but it can be and certainly will be done.[1]

The first transatlantic submarine cable, TAT-8 (Trans-Atlantic Telephone 8) arrived almost exactly on schedule, carrying its first traffic in December 1988. The technology wasn't exactly what Reeves had envisioned. The fibers were much clearer than he had thought would be possible, so the cable needed less than 5 percent as many repeaters as Reeves had proposed. But all in all Reeves was remarkably prescient. Today submarine fiber-optic cables are the international backbone of the global telecommunication network.

The Submarine Cable Business

Submarine telephone cables posed far greater technological challenges than submarine telegraph cables. The first transatlantic telephone cable, TAT-1, was laid in 1955 and 1956, 90 years after the first working transatlantic telegraph cable and began service only a year before Sputnik. Long-distance telephone cables needed amplification, which required vacuum tubes at the bottom of the ocean. Until that technology was mastered, transatlantic telephone calls were sent via radio links, which offered a total of only 15 or 16 channels.[2]

TAT-1 carried 36 telephone circuits on a pair of coaxial cables that ran 3,100 kilometers (1,950 miles) from Newfoundland to Scotland with 51 vacuum tube repeaters each. It cost about a million dollars per two-way channel and was used until 1979.[3] More cables followed, with fatter coaxial cables that carried more channels but transmitted at higher frequencies, where higher attenuation required more closely spaced repeaters. (Transistorized repeaters did not arrive until 1968.)[4]

The practical limit came with the TAT-6, which carried 4,000 voice circuits when completed in 1976 and cost an international consortium $179 million. The TAT consortium immediately started work on a duplicate, called TAT-7, to add 4,000 more circuits in 1983.[5] AT&T designed a new coaxial cable to transmit 16,000 voice circuits, but it would have needed a thousand repeaters to cross the Atlantic,[6] and engineers worried that they couldn't fit the massive cable on a single cable ship and that it would be easily damaged when laid.

Meanwhile, the space age had brought competition in the form of communications satellites. Bell Labs had helped develop the technology, but telephone system operators had been shut out of the satellite business in the 1960s. If they wanted to keep their transatlantic transmission business, they had to find a new form of cable.

Cable operators also could point to significant limitations of satellites. It took a quarter of a second for signals to make the round trip to and from a geostationary orbit 36,000 kilometers (22,000 miles) above the Earth. One bounce is enough to throw off the verbal timing of a conversation, and a second satellite bounce creates more problems. Transmission quality of the analog systems could be erratic, with occasional strong echoes, screeching feedback, or dead-air calls. Satellite channels also lacked security, and that worried security agencies during the Cold War. Citing these problems, AT&T, the British Post Office, and Nippon Telegraph and Telephone (NTT) sought new undersea cable technology.

Allure of Optical Fiber

By the mid-1970s fiber optics was the only viable alternative for undersea cable. The technology had evolved considerably in a decade, with fiber loss reduced to only a few decibels per kilometer, allowing repeater spacing of around 10 kilometers, rather than the 2 to 3 kilometers that Reeves had expected. Engineers could adapt existing submarine cable structures to accommodate fibers instead of coaxial cables. Electro-optic repeaters could fit into the same pressure-resistant housings as the repeaters used with coaxial cables. However, major technological barriers remained.

One was the transmission capacity of the fiber itself. Developers of the first generation of fiber-optic telephone links had focused on short cables linking telephone switching offices. Those systems transmitted light at 850 nanometers through fibers with comparatively large "graded-index" cores 50 or 62.5 micrometers in diameter and could carry 45 million bits per second (672 digitized phone calls) about 10 kilometers. At that distance, however, the system's speed was limited by pulse spreading, which depended on both the type of fiber and the wavelength. Neither the speed nor the distance between repeaters were good enough for submarine cables.

Fortunately two key breakthroughs opened up new possibilities as submarine cable developers started to look at optical fibers. One was the development by J. Jim Hsieh at the MIT Lincoln Laboratory of a new type of semiconductor laser emitting at 1,300 to 1,550 nanometers. The other was the development by Masaharu Horiguchi at NTT of optical fibers that opened new transmission windows at 1,300 and 1,550 nanometers. The new fibers were much more transparent at the long wavelengths than at 850 nanometers, so light could travel several times further through them. In addition, the light pulses did not stretch out as quickly at these wavelengths, so graded-index fibers could carry more bits per second.

Yet submarine cable developers needed the utmost in transmission capacity, so they decided to push for more by returning to the type of fiber Kao had originally proposed. It had a small core—only a few micrometers across—that could transmit light waves only in a single mode rather than in the multiple modes of the larger-core graded-index fiber. Single-mode fiber had been abandoned for use on land in the early 1970s because developers thought it was too hard to align the tiny cores. Yet submarine cablers were willing to take the risk. The next generation of submarine cables was not scheduled until the late 1980s; without the best possible fibers, it might never come. When the time came to make a final decision on submarine cable technology in 1978, AT&T, the British Post Office, and Standard Telephones and Cables became the first developers to commit to using single-mode fiber for any real system.[7]

Serious Single-Mode Development

Bell Labs put transplanted German engineer Peter Runge in charge of developing submarine fiber cables, and cautiously waited a few months before pulling the plug on development of coaxial submarine cables. Some engineers doubted a fiber system would be ready by the TAT-8 target date of 1988, but fiber was the last chance for submarine cables.

Test submarine fiber cables were built around a copper-clad steel "king wire" to provide essential strength. Up to a dozen optical fibers were embedded in a soft plastic shielding layer, gently wound round and round the king wire in a helix. Nylon covered the plastic, forming a core 2.6 millimeters (0.1 inch) thick. Then heavy steel strands were wound around the core, strengthening and shielding the fiber-optic heart of the cable. A welded copper tube covered the steel wires, its role being to carry electric current to repeaters across the ocean. A thick layer of solid white polyethylene covered the copper, making a 21-millimeter (0.827-inch) cable ready to lay on the sea floor like a thick, fat garden hose. Extra heavy steel wires were wound around cable to be laid in shallow water, where fishing trawlers might drag their lines across it. Tests of 100-meter (330-foot) samples began in September 1979,[8] in an artificial ocean Bell had built at its Holmdel, New Jersey, lab to test the coaxial cables for TAT-6. Elaborate instruments controlled temperature, pressure, and cable tension.[9] The results looked good.

Standard Telephone and Cables took undersea cables even more seriously because they were a larger part of the company's business. A major worry of Standard Telephone Laboratories (STL) was that high pressure at the bottom of the ocean would cause the fiber to wrinkle inside the cable, causing "microbends" that would let light leak out. To assess the problem, they designed a cable with four graded-index and two single-mode fibers inside an inner aluminum tube. They assembled 9.5 kilometers (5.9 miles) of cable, which the Post Office carefully laid in Loch Fyne, Scotland, in February 1980. Later they added a repeater capable of transmitting 140 million bits per second. Careful measurements showed that the cable worked well in the water, and bolstered the case for single-mode fiber.[10]

By 1980 the course was clear. Peter Runge of AT&T Submarine Systems told a laser conference, "The next generation of coaxial cables will not be developed because of fibers." When TAT-7 was laid in 1983, it would be the last coaxial cable to cross the Atlantic. The next transatlantic cable, TAT-8, would contain two pairs of single-mode fibers carrying 280 million bits per second, the combined equivalent of 35,000 phone calls—nearly nine times the capacity of TAT-7. Plans called for repeaters to be 30 to 35 kilometers (19 to 22 miles) apart. He said the cost per channel should be only 20 percent that of TAT-6 and half the cost of the coaxial system Bell had abandoned as impractical.[11]

Bell had its single-mode fiber research group at Murray Hill make 22 single-mode fibers, each 20 kilometers (12 miles) long, for a cable to be tested in the North Atlantic. They were first used in some experiments at Holmdel, then shipped to New Hampshire, where Simplex Wire and Cable put a dozen fibers into an 18-kilometer cable. In September 1982 an AT&T cable ship laid part of the cable and a repeater 5.5 kilometers (3.4 miles) under water on the Atlantic floor. It was the first fiber cable tested at that depth, and it worked perfectly in the water.[12]

By mid-1982, however, long-term tests were showing some distressing trends. NTT had laid a couple of cables for field trials in 1979, then cautiously checked back every several months to monitor aging. In June 1982 they discovered the loss in the whole fiber was increasing at 1,300 nanometers.[13] STL saw the same pattern at Loch Fyne, although it was not as severe.

Uneasy rumors that something mysterious was amiss spread through the fiber optics world. Uneasy engineers went back and measured their cables. Many found no problems, but some noticed increasing loss. Increasing fiber loss knocked out a link operated by Britain's Central Electricity Generating Board. The problem "hit like a sledge hammer" in Britain, recalls John Midwinter, then head of fiber optics development at British Telecom Research Laboratories.[14] The Japanese were even more anxious; they had seen a cable go totally dark.[15] Developers of submarine cables had put all their eggs in the fiber-optic basket, and it looked like they were in danger of breaking.

The first crucial clues came from a Japanese test of the long-held assumption that water did not affect fiber transmission. In early 1982 Naoshi Uesugi of NTT filled a cable with water and put an uncabled fiber in the same water to compare what happened to transmission. After eight months, loss of the cabled fiber increased dramatically at 1,080 to 1,240 nanometers and beyond 1,400 nanometers. Attenuation of the uncabled fiber had hardly changed.[16] The affected wavelengths indicated that hydrogen was absorbing light. Somehow the affected fibers were soaking it up. Yet the changes were inconsistent; they affected some fibers and some cable structures but not others.

Developers in Japan, America, and Britain set up task forces to attack the problem. Once they focused on hydrogen, they began to see patterns. The only fibers affected were those containing phosphorous, which softened the glass, making it easier to draw. British work showed that if phosphorous made up about 1 percent of the glass it reacted with hydrogen, forming chemical bonds that absorbed light.

Differences also showed up in cable structures. The high voltages carried by wires to power undersea repeaters caused metal in the cable to react with the traces of water that inevitably seeped into the cable.[17] The high voltages broke down the water to yield hydrogen that could accumulate in the cable. The extent of the problem depended on where the hydrogen collected. STL's original design had a problem because an internal barrier trapped hydrogen in the same region as the fiber.[18]

With the culprit identified, fiber developers abandoned phosphorous and added hydrogen-resistant coatings, while cable manufacturers added hydrogen seals, eliminated plastics that might release hydrogen, and sealed their cables tighter to keep water out.[19] Careful measurements of the new cables showed the problem was fixed, and developers pushed onward with submarine fiber cables.

A Go-ahead for Submarine Cables

In December 1982 an international consortium of companies[20] requested bids for the TAT-8 cable running from Tuckerton, New Jersey, to a point off the European coast, where it split into separate cables to France and England. AT&T, Standard Telephones and Cables, and the French Submarcom consortium were the only bidders. AT&T tried to land the entire project, but the cozy little world of international telephony usually split contracts to assuage politicians and corporations. In the end, AT&T got the lion's share of the $335 million contract, from America to the branch point. STL got the British section, and Submarcom the French part.[21]

A transpacific consortium was just months behind TAT-8, with service to start in 1989. Cables had lagged far behind satellites across the wider Pacific. Submarine cables carried just under a thousand voice circuits from California to Japan in 1984; 1,460 more went from Canada to Australia via Hawaii.[22] AT&T landed a contract to build the Hawaii 4 cable from California to Hawaii and the part of Trans-Pacific Cable 3 from Hawaii to a branching point off Japan. The Japanese Ocean Cable consortium won the contract to build the rest of the system.

Meanwhile, work on smaller submarine cables went much faster. For each massive cable like TAT-8, there were dozens of short ones, typically linking offshore islands to the mainland or other islands, or linking major coastal cities. The high capacity and long repeater spacing possible with fibers also were important for many of these systems. British Telecom (the former telecommunications branch of the Post Office) was the first to use a fiber cable for a short submarine link carrying regular traffic. In 1984 it laid an 8-kilometer (5-mile) fiber cable from the coastal city of Portsmouth to the Isle of Wight in the English Channel. Standard Telephones and Cables supplied the hardware, which carried 140 million bits per second over each of four pairs of single-mode fibers. The whole system cost just $292,000 and contained no repeaters.[23]

More important was the first fiber-optic addition to the array of cables crossing the English Channel, UK-Belgium 5. Traffic was growing between Britain and Belgium, Germany, and the Netherlands; the carriers needed more capacity, and they wanted digital transmission. STL won the job by offering to lay a "fully-engineered prototype" that would start with experimental service, then phase into regular use. A dozen cables already ran from Britain to Belgium and the Netherlands, carrying just over 23,000 voice circuits.[24] The fiber cable carried 11,500 on three pairs of single-mode fibers. The three repeaters in the 122-kilometer (76-mile) cable were the first on a submarine fiber cable to carry regular telephone traffic.[25]

Laying the main part of the cable took just five days in early 1986, although it was buried in the seabed to prevent damage from trawlers and ship's anchors. A second cable ship laid the rest of the cable in shallower water, where a submersible trencher buried it deeper in the sea bed. Tests began as soon as engineers hooked up the cable; commercial service formally began on October 30, 1986, with a two-way video conference between London and Ostend, Belgium.[26]

The Japanese kept pace. NTT designed a system to transmit 400 million bits per second between islands in the Japanese archipelago. After two years of shallow-water tests, in November 1984 they laid 28 kilometers (17 miles) of cable and two repeaters in the deepest water handy, the 8-kilometer-deep Ogasawara Trench. Careful tests showed little change, so reassured NTT engineers went ahead with plans for a submarine cable between Japan's two main islands.[27]

Shark Attack in the Canary Islands

With the lion's share of the two biggest submarine fiber cables, AT&T carefully planned a test of TAT-8 technology in the Canary Islands off Africa, where the Spanish telephone authority wanted a cable between Grand Canary and Tenerife. Canary Islanders got the latest submarine cable technology in return for putting up with exhaustive tests of the hardware, including trials of recovery of the cable from the ocean floor and shipboard repairs.

AT&T got more than it bargained for. Installation of the 119-kilometer (74-mile) cable went well, including a pair of repeaters 40 kilometers (25 miles) apart. Soon afterward, though, engineers noted a short circuit in the cable, when ocean water reached the central conductor that carried electric power for the repeaters. The AT&T team worried about how water was getting into the cable, but it wasn't an immediate catastrophe because the cable was designed to withstand a single short. Then a second short followed, knocking out repeater power and forcing AT&T to haul the cable up from the ocean floor.

Examining the recovered cable, AT&T engineers found shark teeth embedded in the outer plastic coating. Sharks had bitten the cable, and left some of their easily detached teeth behind. The engineers were amazed; sharks had not been known to bite undersea coaxial cables. Marine biologists also were amazed; they had not known that sharks swam deep enough to reach a cable more than a kilometer (0.6 mile) down. News inevitably leaked out, and the press had a field day.

Sharks had ignored coaxial cables. Why should they turn their razor-sharp teeth on fiber? Biologists stretched a cable across a tank, and captive sharks ignored it until they switched on the current that normally powered repeaters. The electric field attracted the predators, who apparently use it in hunting. Coaxial cable also carried a current, but its outer metal wrap blocked the electric field. Nothing blocked the field from reaching the water around the fiber cable, where the sharks could sense it.[28] Other factors also may have played a role. The Canaries cable hung above the ocean floor in some areas and also made vibrations, which may have attracted sharks—or perhaps the cable simply got in their way.

Engineers took no chances when they returned the cable to the ocean floor, wrapping the cable with a strong steel tape. They also extended the armor normally used to protect the cable in shallow water to depths well below where the sharks had bitten the cable. As they laid the cable, they made sure it rested on the bottom rather than hung in the water where it might tempt hungry sharks. No shark bites have been reported since then.[29]

The reliability of semiconductor lasers initially looked like a more serious problem. Dreading the cost of hauling the cable from the ocean floor to replace a dead laser, engineers designed repeaters with three spare lasers to back up each laser transmitting signals.[30] However, the lasers improved so much as time passed that AT&T's final TAT-8 repeaters included just one backup laser per fiber.[31] The transmitters, cable, and receivers also worked better than had been expected in 1980, allowing AT&T to double repeater spacing from the planned 30 kilometers to beyond 65 kilometers. STL and

Submarcom were more cautious, spacing repeaters about 40 kilometers in their parts of TAT-8.[32]

Turning the Tables on Satellites

In the mid-1970s, the communications satellite industry thought it might vanquish costly cables from the transatlantic market.[33] A decade later, fiber optics had turned the tables.[34] As TAT-8 moved toward completion, international consortia planned more submarine cables. In May 1986 AT&T, British Telecom, Teleglobe Canada, and the French and Spanish telephone authorities agreed to plan and build TAT-9, with service to start in 1991.

The new cable was designed around a fourth generation of fiber technology, operating at 1.55 micrometers, where fiber loss was lowest. Both Corning and British Telecom were pushing the advantages of fibers they had developed with minimum pulse dispersion shifted to 1.55 micrometers.[35] AT&T, however, chose instead to use lasers that emitted an extremely narrow range of wavelengths, so pulses should suffer little dispersion at 1.55 micrometers. Their design let them double transmission speed to 560 million bits per second and stretch repeater spacing to 100 kilometers in TAT-9. The same technology later reached 140 kilometers.[36]

The satellite industry fought back. The International Telecommunications Satellite Organization (INTELSAT) promised to cut its rates and claimed new echo-cancellation circuits made complaints about satellite delays "spurious."[37] (As a frequent international caller, I knew better; satellites still had delays and bad circuits.) Each side claimed its service was cheaper, but satellites were in retreat. As plans for new cables proliferated, the satellite industry quietly turned to routes not well served by fiber cables and services other than telephony, including video and data communications.

TAT-8 planners came remarkably close to their original plans of starting operation in July 1988. Ironically, the delays were due to relatively mundane electronic components, not fibers, cable, or lasers. In August, AT&T's cable ship *Long Lines* laid its portion of the cable, stopping at the branch point off the French coast. French and British cable ships met it and laid the rest of the cable from the branch point to land.[38] After a few months of testing, the cable went online in late December,[39] just beating Alec Reeves's prediction of two decades earlier.

The job was not finished. Strong currents off the French coast exposed the buried cable, and trawlers snagged it in February and March, damaging the wires carrying repeater power. In March an electrical fault knocked out the British leg of the cable. Calls had to be rerouted via satellite until repairs were finished and the cable was buried deeper in early April.[40] Meanwhile, the transpacific cable began service to Japan.[41]

The new cables did not revolutionize international communications overnight, but I could see the improvement. My calls to London were more likely to go through at the peak hours for transatlantic business. I got fewer bad circuits with maddening echoes or dead air from England. More and more calls were as clear as any long-distance call made over fiber-optic lines in America. The most reliable sign of a transatlantic call became an English accent on the other end.

Optical Amplification and the Telecommunications Boom

Better technology brought higher cable capacity at lower cost per channel. The demand for international telecommunications exploded, filling TAT-8 to capacity soon after it began service. TAT-10 and TAT-11 followed with the same 1.55-micrometer technology as TAT-9. But the real boom was yet to come.

In 1963 Elias Snitzer[42] showed that a laser without mirrors at either end could amplify a light signal passing through it. Over the years Alec Reeves and a number of other telecommunications developers envisioned using optical amplifiers in fiber-optic systems, but no one was sure how to make a practical version. In the end, the winner proved to be a variation on Snitzer's original idea, doping small amounts of a light-emitting element into the core of an optical fiber and exciting the atoms with light from an external source. In the late 1980s, David Payne of the University of Southampton found the right element, erbium, which nicely amplifies light at the 1,550-nanometer wavelength where optical fibers have their lowest loss.[43] Emmanuel Desurvire at Bell Labs and others helped develop the technology into a practical optical amplifier.

Submarine fiber-optic cable developers adapted erbium-doped fiber amplifiers at record speed. They were nearly ideal for submarine cables because they could amplify optical signals directly, replacing more complex repeaters that converted the light signals into electronic form for amplification then had to convert them back to light for transmission through the fiber. Fiber amplifiers easily fit into the torpedo-shaped cases designed to house undersea repeaters, and by the mid-1990s they were rolling off production lines as part of new cables that could carry signals at rates to 5 gigabits per second—the equivalent of 64,000 telephone calls.

Meanwhile, researchers showed that fiber-optic cables equipped with erbium-doped fiber amplifiers could transmit separate signals simultaneously on different wavelengths, a technique called wavelength division multiplexing. By the late 1990s each fiber in a submarine cable could carry signals on 4 to 8 separate wavelengths, or optical channels. The limit came not from the inherent capacity of the fiber but from the need to balance carefully the gain of each amplifier on each optical channel and to supply the electrical power needed to drive the lasers that energized the erbium atoms in the fiber amplifier. Shorter cables can carry dozens of wavelengths, and laboratory demonstrations have carried more than 100 wavelengths per fiber.

That tremendous potential stimulated a sudden boom in construction of submarine fiber-optic systems. Several companies set out to build global networks of submarine cable to meet what they thought was an insatiable demand for international transmission capacity that came with the Internet boom. Some, like Project Oxygen, never raised the money they needed, but others did, investing billions of dollars in vast networks. As one prophetic study predicted in the 1980s, the ocean floors were covered with glass. Then the Internet bubble burst.

Today the ocean floors are still covered with glass, but the fiber networks are far beyond today's needs. Global Crossing, which built the biggest network, is in bankruptcy, and some others are shaky. Investors have lost billions of dollars. Yet like the burst of railroad construction in the late nineteenth century that ended in a financial collapse, the fiber-optic boom leaves behind a legacy of infrastructure available for the future.

Notes

This material is adapted, with permission, from Jeff Hecht, *City of Light: The Story of Fiber Optics* (New York: Oxford University Press, 1999). The quotation in my title comes from a now-forgotten source that predicted strong growth for undersea fiber-optic cables.

1. Alec H. Reeves, "The Future of Telecommunications: Bernard Price Memorial Lecture 1969," *Transactions of the South African Institute of Electrical Engineers* 60 (September 1970), 445–465.

2. Institution of Electrical Engineers, "Sailing Ship to Satellite: The Transatlantic Connection," exhibit brochure (London, 1994). The longwave band allowed one channel; shortwaves allowed fifteen.

3. "Submarine Cable Systems," in E. F. O'Neill, ed., *A History of Engineering and Science in the Bell System*, vol. 7, *Transmission Technology (1925–1975)* (Indianapolis: AT&T Bell Laboratories, 1985), 337–371.

4. Early germanium transistors were short-lived and much more sensitive to temperature than modern silicon transistors, so there was some reason for the delay. R. D. Ehrbar,

"Undersea Cables for Telephony," in *Undersea Lightwave Communications*, ed. Peter K. Runge and Patrick R. Trischitta (New York: IEEE Press, 1986), 3–22.

5. Richard J. O'Rorke, Jr., *1990 World's Submarine Telephone Cable Systems* (Washington: Government Printing Office, 1991).

6. I use only statute (land) miles here; repeater spacing is often given in slightly longer nautical miles. The spacing for TAT-6 was 2.5 nautical miles. Ehrbar, "Undersea Cables for Telephony," 3–22.

7. Peter Runge, interview, July 19, 1995.

8. Peter Runge, "A High-Capacity Optical-Fiber Undersea Cable System" (paper presented at Conference on Laser and Electro-Optic Systems, San Diego, 1980).

9. O'Neill, "Submarine Cable Systems."

10. "Britain and an ITT Subsidiary Begin Tests of Undersea Fiber Optic Cable," *Laser Focus* 16 (May 1980), 66–68; quotes article by P. Worthington, "Application of Optical Fiber System in Undersea Service," in *Institution of Electrical Engineers, International Conference on Submarine Telecommunication Systems* (February 1980), 26–29.

11. A third pair of single-mode fibers was included in the cable in case of fiber failure. That caution was reflected in the design of the entire TAT-8 system. "Bell Plans a Single Mode Fiber System for Transatlantic Installation by 1990," *Laser Focus* 16 (May 1980), 68–70.

12. Peter K. Runge, "Deep-Sea Trial of an Undersea Lightwave System," in *Technical Digest, Topical Meeting on Optical Fiber Communications* (Washington, D.C.: Optical Society of America, 1983), paper MD2.

13. Naoya Uchida and Naoshi Uesuge, "Infrared Optical Loss Increase in Silica Fibers Due to Hydrogen," *Journal of Lightwave Technology* LT-4 (August 1986): 1132–1137.

14. John Midwinter, interview, December 5, 1995.

15. Murray Ramsay, interview, December 2, 1994.

16. Uchida and Uesuge, "Infrared Optical Loss."

17. Kiyofumi Mochizui et al., "Influence of Hydrogen on Optical Fiber Loss in Submarine Cables," in Runge and Trischitta, *Undersea Lightwave Communications*, 177–188.

18. Murray Ramsay and Martin Chown, interview, December 12, 1994.

19. Mochizui et al., "Influence of Hydrogen on Optical Fiber Loss in Submarine Cables."

20. The number quoted varies; an initial AT&T press release (May 16, 1983) says that 29 companies are involved; a directory of submarine cables lists 23.

21. AT&T press release, dated November 16, 1983.

22. "World Submarine Telecommunication Systems," map dated September 1984, published by STC Submarine Cables Ltd.

23. O'Rorke, *1990 World's Submarine Telephone Cable Systems*; it lists only 1.3-micrometer transmission.

24. Release supplied by Nortel, dated October 30, 1986, from their files.

25. George A. Heath and Martin Chown, "The UK-Belgium No. 5 Optical Fiber Submarine System," in Runge and Trischitta, *Undersea Lightwave Communications*, 129–141.

26. Release supplied by Nortel.

27. Hiroshi Fukinuki et al., "The FS-400M Submarine System," in Runge and Trischitta, *Undersea Lightwave Communications*, 69–82.

28. Runge interview.

29. Runge interview.

30. Runge, "A High-Capacity Optical-Fiber Undersea Cable System."

31. Fridolin Bosch, G. M. Palmer, Charles D. Sallada, and C. Burke Swan, "Compact 1.3:m Laser Transmitter for the SL Undersea Lightwave System," in Runge and Trischitta, *Undersea Lightwave Communications*, 445–458.

32. O'Rorke, *1990 World's Submarine Telephone Cable Systems.*

33. See, for example, Michael Kinsley, "Is AT&T Hamstringing Comsat," *New York Times*, June 13, 1976.

34. R. A. Williamson, project director, International Cooperation and Competition in Civilian Space Activities, OTA-ISA-239 (Office of Technology Assessment, July 1985).

35. The technology for dispersion-shifted fibers was refined gradually by fine-tuning the structure of the fiber. V.A. Bhagavatula, M. S. Spotz, W. F. Love, and Donald B. Keck, "Segmented-Core Single-Mode Fibers with Low Loss and Low Dispersion," *Electronics Letters* 19, no. 9 (April 25, 1981): 317–318; B. J. Ainslie et al., "Monomode Fibre with Ultra Low Loss and Minimum Dispersion at 1.55 Microns," *Electronics Letters* 18 (1982): 842–844.

36. Runge interview.

37. Leonard Heymann, "Intelsat Mounts Defensive Campaign against Cables," *Communications Week*, September 22, 1986, 51.

38. Kathleen Killett, "Fiber Optic Cable Approaches Transatlantic Service Cutover," *Communications Week*, December 5, 1988, 32.

39. David Zielenziger, "Trans-Atlantic Link Goes on Call," *Electronic Engineering Times*, December 26, 1988, 16.

40. Notes by author for story written for *New Scientist*, April 6, 1989.

41. Karen Lynch, "Fiber Link to Japan Complete," *Communications Week*, April 24, 1989, 8.

42. Charles J. Koester and Elias Snitzer, "Amplification in a Fiber Laser," *Applied Optics* 3, no. 10 (October 1964): 1182–1186.

43. R. J. Mears et al., "Low-Noise Erbium-Doped Fiber Amplifier Operating at 1.54 μm," *Electronics Letters* 23 (1987): 1026; also interview with David Hanna, May 22, 1995.

Managing the Network

5

The International Telegraph Union: The Cable Companies and the Governments

Jorma Ahvenainen

Putting the Question

This paper examines how the telegraph, introduced in the 1850s as a means of exchanging official information, had by the outbreak of the First World War come into general public use all over the world. Various factors are at play here. Many are technical issues pertaining to the lines and cables themselves; nevertheless, the development of the telegraph depended not only on how the lines and cables were laid but also on the pricing of the telegraph services. The first lines within Europe, and in many non-European countries too, were intended solely for government use, something like an official postal system, the use of which could not really be said to have any particular price.[1] Notable exceptions were Britain and the United States, where the telegraph was introduced for public use from the very start. Although governments generally realized the potential value of the telegraph for public use, there were also misgivings about making it publicly available. The telegraph was being developed in Europe at precisely the same time as the revolutionary ideas of the "mad year" of 1848 were fomenting popular unrest across the continent. Some feared that uncontrolled use of the telegraph would encourage the even swifter spread of subversive ideas.

Governments tried to prevent the telegraphing of information that might undermine their security or pose a threat to public order by instructing telegraph officials to monitor the contents of the messages they handled. Another way of controlling the use of the telegraph was through the tariffs. When private correspondence came to be permitted, tariffs were set so high that use of the system was a luxury that only the wealthiest could afford. An example of this is provided by France's earliest tariffs: in the year when the French telegraph was made available to private correspondents, the price of a 20-word message (the shortest allowed) between Calais and Marseilles was 17 francs, equivalent to about fifteen days' pay for a manual worker of the time.[2]

From the point of view of worldwide communications, intercontinental submarine cables were at least as important as landlines. This is not the place to describe the early history of the cables. Suffice it to say that the first underwater cable was laid between Dover and Calais in 1851, and the first functioning cable between Europe and North America came into use in 1866. Europe was connected with the Far East and Australia in the first half of the 1870s, partly by means of cables via India and partly via Siberia's landlines and cables between Russia, Japan, and China. The cable link between Europe and South America was opened in the middle of the 1870s, and that between North and South America came into service in 1882. The last major underwater cables were laid beneath the Pacific in the early years of the twentieth century, to link North America with East Asia. With only one exception, these underwater cables were laid and operated by private companies. In 1914 private companies owned a total of 236,000 nautical miles of submarine telegraph cables, while government-owned cables amounted to only 47,000 miles.[3] To determine how the telegraph came into common use, we need to consider the origins and operations of the International Telegraph Union, the cable companies, and government policy toward the companies.

Origins of the International Telegraph Union

The International Telegraph Union (ITU) owed its origins to two sets of circumstances. In 1849 Prussia and Austria signed an agreement allowing telegraph traffic between Berlin and Vienna. Prussia had taken the initiative in this case, and it took steps to try to reach similar agreements with other German states. From these originally bilateral agreements emerged a larger organization, the German-Austrian Telegraph Union (Deutsch-Österreichischer Telegraph Verein), established in Dresden in 1850. Its founding members were Prussia, Austria, Bavaria, and Germany, but other German states and also the Netherlands joined in due course. The aim of the union was to provide both official and private users of the telegraph with the advantages of a system based on one common set of rules and regulations. Governments that joined the union agreed to coordinate their telegraph legislation and to standardize tariffs. The agreement also called for occasional meetings of the signatories.[4]

At one such meeting, held in Vienna in 1851, decisions were made that would prove to be important for the subsequent development of international telegraph traffic. Officials determined that it should be made possible to send telegraph messages from one country to another without their being stopped at the border and then sent on again. Another established principle was that everyone had the right to use the telegraph and that secrecy should be guaranteed. The provider of the service had an obligation to transmit messages quickly and accurately, and technical standardization meant that Morse's telegraph machine was approved as the only one to be used for international traffic.[5] At the meeting in Stuttgart in 1857 the most important decisions concerned regulations, which were divided into two parts: matters connected to the basic principles of the union were allocated to the so-called convention side, while matters that affected the everyday running of the telegraph and were subject to change were included in the so-called regulations, or operations, instructions.[6]

A similar development to that taking place under the German-Austrian agreement was occurring in western Europe. At the beginning of the 1850s France had begun to negotiate agreements with its neighbors to regulate the use of the telegraph. In 1855 an agreement was signed by France, Belgium, Switzerland, Sardinia, and Spain to set up the West European Telegraph Union, the purpose of which was very similar to that of the German-Austrian Union. Other neighboring countries soon sought to join this union.

The result of these developments was that Europe began to be organized into two separate telegraph groupings. They could not, however, be entirely independent of each other, since simple practicalities demanded that they should reach agreement on basic aspects of telegraph communication. In practice, problems were sorted out in negotiations that covered a wide range of topics. The most important meeting, at the end of the 1850s, took place in Brussels and was attended by Belgium, France, and Prussia. The participants reconsidered earlier regulations governing the telegraph, authorized for the first time the use of numbers in messages and agreed that the maximum permitted length of any single word should be seven letters. Tariffs were reduced by common agreement. This settlement was so attractive that other countries were soon tempted to adhere: Spain, Portugal, Sardinia, Switzerland, and Turkey signed in 1859; the Scandinavian countries, the Vatican, the two Sicilies, and Russia signed in 1860; and Luxembourg signed in 1861. Prussia brought the whole of the German-Austrian Telegraph Union along with it.[7] Thus most of Europe found itself operating under the terms of a single telegraph agreement as early as 1861, creating virtually the same situation as would have held had the various countries officially connected their lines one to another. It must be noted, however, that Britain was not a party to this agreement. The British telegraph service was then privately owned and split up among many different owners: people who could negotiate and sign treaties were difficult to find.

In 1864 the French government invited the most important European nations to take part in an international telegraph congress, to be held in Paris, with the objective of reaching agreement on the establishment of a single telegraph system. Twenty states sent delegates in May 1865.

Behind the French initiative was the assumption that it was up to France, as the foremost nation in Europe, to take the lead in making proposals and trying to organize how things were done. The French government aimed to find a way to unite the two existing European systems into a single organization. The government came to the conference table armed with its proposals for a new treaty and a statute to govern the service, and it carried the day. The outcome of the congress was a convention that included clauses on tariffs as well as regulations for the use of the telegraph. The delegates approved the convention on May 17, 1865.[8] The states that were party to the treaty were all European except for Russia and Turkey, which included territory in Asia too. For the reasons already mentioned, Britain did not attend the congress.

Thus was established the ITU. The successful outcome of the congress was due largely to the fact that the task had been to create one new organization from two preexisting and essentially similar ones. The task was also made easier by the fact that the question was not primarily political but depended rather on technical and practical considerations. Nevertheless, the achievement was a feather in the cap for the French government, which through its Telegraph Administration led telegraph activities in Europe at this time. It was a victory for the French too in that the French language became the official language of the union and international telegraph tariffs were henceforth to be calculated in French francs.

Just as in the two European conventions on the telegraph that had preceded the Paris convention, so the ITU was to hold conferences from time to time to review the agreement and the regulations. The first conference after Paris was held in Vienna in 1868. There the decision, taken in principle in Paris, to allow private companies to join the union was confirmed. Any companies taking advantage of this would of course enjoy the benefits of membership while being obliged to observe the union's rules.[9]

At the next meeting, however, which began in 1871 in Rome, the question of the status of private companies in the union was a key issue. Since the time of the Paris conference the expansion of the cables had been particularly in the private sector, and private companies already operated some intercontinental lines. This situation could not be ignored. Disagreement did, however, arise on what exactly the position of the private cable companies should be in the organization. After protracted negotiations, it was agreed that the companies could be present and could speak at both sectional and plenary meetings, but they would not have voting rights. Britain, which in 1869 had made the country's domestic telegraph system a state monopoly and was now represented in Rome, was especially active in pressing for recognition of the rights of private companies.[10]

Private Cable Companies

Private cable companies were founded in order to run a business, just like any other company. There was no direct government involvement in the laying and ownership of the international submarine cables. If governments had been involved, they would have had to seek authorization for their telegraph stations and cable staff on foreign territory. This being impossible, the ocean cables were laid by private companies, even when they were in fact underwritten by government money. The ocean cables thus had a different point of departure from that of the government-laid landlines or sea cables, which were generally relatively short. Ocean cables required considerable capital investment, and the levels of technology and geographical knowledge prevailing at that time meant that the cable companies ran big risks in undertaking this business. In official terms too the ocean cables were more complicated than landlines in that, running between different countries, they were operating where the entrepreneurs were not on their native soil.

Once a company had obtained the necessary license, it had to raise the funds it needed to realize its project. The capital was generally put forward by private individuals; banks were not interested in subscribing guarantee capital precisely because of the high risk involved. The normal procedure was to found a consortium, in which wealthy individuals subscribed shares and some fractional part was set aside for public subscription. The proposed company was presented at the same time as the share offer and, as one would expect, the prospectus painted a very rosy picture of its future. Otherwise, the whole project would have been pointless.

It was, however, essential that the company make every effort to fulfill its promises, and according to these plans income and expenditure also had to be organized. While the cable itself was a large fixed expense that could not be altered in any way once it had been laid on the sea bed, it was possible to influence running costs. It was necessary to have a sufficient income for both capital and running expenses, so ensuring the generation of adequate receipts was fundamental to the companies' success. The companies therefore had to establish tariffs that were at the same time low enough to attract as much business as possible yet high enough to yield sufficient profit so that obligations could be honored.

When ocean cables were first laid in the 1850s, there were neither national nor international regulations controlling the activity. The companies themselves, however, proceeded to seek licenses from the governments of countries where they wanted to bring their cables ashore and operate services. Governments then began to grant concessions for the telegraph in their territory even when the end of the cable had been brought onto privately owned land.

Countries that occupied a strategic geographical position bordering the oceans and yet did not have their own cable companies gave concessions to foreign companies. Within Europe, Spain is a good example of this. Despite being a colonial power, Spain had neither the technical knowledge nor the necessary financial resources needed for cable operations. Accordingly, in the mid-1850s the Spanish government began to grant concessions to foreign companies in order to establish connections to South America and the Caribbean, and from there to North America.[11] In return for the concessions the government was entitled to free official use of the network. This was of no little importance to the Spanish government, since its colonies were located in the very regions through which the cables had to run. In the Far East, at the beginning of the 1870s the Japanese government granted the Great Northern Telegraph Company the right to lay a cable connecting the island state to the mainland of Asia and gave the same company a long-term monopoly in Japanese international telegraph traffic.[12] The British government also granted cable rights to non-British companies in its colonies.[13]

The laying of cables beneath the oceans and between continents took place in what is known as the age of liberalism. Nonetheless, the underwater cable business by no means reflected the liberal spirit of the time. In seeking government licenses, companies asked for exclusive rights; otherwise, the cables would probably have been out of the question. The cable and all the equipment were so expensive that it was vital to be able to depend on a certain amount of business, without competition over prices. This put governments into a difficult position: the cables they needed could become available only if they granted a monopoly. The British Board of Trade did indeed try to prevent the cession of exclusive rights on British territory, but it was not altogether successful in this. Even if the government in London was not in favor of monopolies, the governments of the dominions themselves considered the cable connections to be so important that they were indeed prepared to go to the length of granting exclusive rights, if that was the price that had to be paid for getting them. In 1854 the Atlantic Telegraph Company, in bringing a cable to British Newfoundland, was granted exclusive rights. But when the cable soon failed, the company lost its monopoly, and after that no exclusive rights were granted in the North Atlantic area. The competition became free.[14] Sometimes governments tried to limit the effects of the monopoly by setting maximum tariffs when they granted the privilege. When in 1866 the United States Congress granted the first cable concession in the country, to operate between Florida and Cuba, it set a maximum tariff of $3.50 for ten words.[15] When in 1864 the Italian-born international businessman Alberto Balestrini sought and obtained recognition at the Paris international congress for his far-reaching monopoly position in the South Atlantic, it was laid down in the same document that in any negotiations with Balestrini governments had the right to determine the price of messages.[16] Balestrini forfeited his concession by failing to meet deadlines, but when the Portuguese government granted a new concession to a British firm, the Telegraph Construction and Maintenance Company, in 1872, it also fixed the tariff between Portugal and Brazil at a maximum of F 140 for a 20-word message.[17]

In many cases, however, not even a monopoly was sufficient to guarantee the viability of the telegraph service. It was difficult to make a profit when the cable was in areas with relatively few inhabitants, remote from the center of things—that is, remote from Europe. Typically, either the company or representatives of the local population would propose that cable service should be underwritten by government subvention. This pressure was particularly effective in the colonies—first British and then those of other European countries—where people wanted to have a rapid means of communicating with their former homes. The existence of the cable network was also important to the British trading houses. So it is not at all surprising that these big business magnates were major shareholders in the cable companies.

After the failures of some of the cable companies in the 1850s, the British government was extremely unwilling to give them any direct financial support. Only in rare cases of evidently exceptional political or military significance was support given. The clearest example was in the establishment of a cable connection between Europe and the Cape. The link was made from Aden via Zanzibar to Cape Town in 1879 with the help of London, the two colonial governments in the Cape and Natal, and the Portuguese government. In Aden the cable was linked up to the eastbound cable of the Eastern Telegraph Company.[18] When, however, the cable companies in the Caribbean were facing hard times in the 1880s and tried to persuade the British government to bail them out, the government made it clear that if the Caribbean sugar plantation owners and traders could afford to do without the telegraph, so could London.[19]

Whoever paid the subventions, the situation tended soon to become monopolistic. Once a subvention had been granted to a company for a year or perhaps even for ten years, the door was then closed to other entrepreneurs, unless the first grantee made a really bad job of things.

Cable Companies and the International Telegraph Union

The general meeting of the union in St. Petersburg in June–July 1875 was an important one from the point of view of international cooperation over the telegraph, because the delegates decided to establish a completely new convention. It was far more straightforward than the earlier ones, with only 21 articles. The convention was intended to be permanent, unlike regulations and tariffs, which were to be revised from time to time. Britain was now also present, and it agreed to abide by the telegraph regulations and tariffs, although it still refrained from signing the agreement. As for private companies joining the union, it was stated briefly that they could be approved on the recommendation of the country in which the union's conference had last been held. Any private companies that did join were to be afforded the advantages brought by the agreement and had to accept the obligations that followed (Article 83).[20]

At the time of the St. Petersburg conference very few private companies were members of the union, but afterward the number signing the treaty and its appendices grew. By 1914, 19 private companies were members of the union, and there was a group of 15 companies that were not official members but nevertheless respected the union's regulations for the functioning of the telegraph and were correspondence clerks of the union. In addition there were a few cable companies that were linked to the union through some other company.[21] For companies the international treaty was important in the following ways:

- It established the principle that everybody had the right to use the telegraph; indeed, since the telegraphs were a business, it would have been absurd for the companies to refuse to accept messages. Even so, the cable companies too had a serious problem in some of the less developed countries with the occurrence of insurrections or revolutions, during which times the companies' lines and other property were deliberately damaged.
- The international treaty required that telegrams should be secret, and observance of this requirement was generally easy for the companies. It was another question, however, how far the companies could protect this secrecy when a government had declared a period of censorship.
- It was of crucial importance to the cable companies that there should be regulations governing the activities of the telegraph. As the business was

spread all over the world, serious difficulties would have arisen if countries and states had operated according to their own rules.

• The tariffs set down by the union divided countries into two groups: those within Europe (*Régime Européen*) and those outside Europe (*Régime Extra-Européen*). At the time of the St Petersburg treaty, all the members of the union were Europeans, but some of the members brought territories outside Europe into the tariff agreement: Britain brought India; France brought Algeria, Tunisia, and Indochina; and Russia and Turkey brought their Asian lands. The signatories to the agreement had approved tariffs for Europe and for outside Europe, composed of two parts: an office tax (*taxe terminale*) and a transit tax (*taxe de transit*). The telegraph offices of the countries in which the telegram was sent and received levied the office tax as their payment. The transit tax was levied by the country or countries through which the telegram was sent, and it was the payment for the use of the lines.

An example is useful: when a telegram from Germany had to be connected in Lisbon to the South American cable that began there, the costs in 1879, when the cable connection opened, were as shown in table 5.1.[22] In this example the office tax in the country of final destination of the telegram is not included, because in 1879 the South American countries were not part of the ITU. In fact, the terminals of the telegraph in South America usually belonged to the cable companies themselves, so it was not uncommon for telegrams to be free of a final office tax there. To complete the example given above, the price for sending the message on from Lisbon to Buenos Aires, for instance, would have been F 18.955, so the total price from Germany to Buenos Aires was F 19.705 per word.[23] Brazil was in fact the first South American country to join the union, in 1877, and it was followed by Argentina in 1889, Uruguay in 1902, Bolivia in 1907, and Chile in 1908.

For the cable companies it was particularly important that the European countries should have agreed among themselves about the rates that were to be charged to the point where the cables left European shores. The tariffs for

Table 5.1
Taxes, in Francs per Word, between Germany and Cable at Lisbon

Germany, office tax	0.2250
France, transit tax	0.2250
Spain, transit tax	0.1875
Portugal, transit tax to cable	0.1125
Total	0.75

most cables within Europe were calculated from London, Brest, Le Havre, Lisbon, and Cadiz. The European lines from inland places to the coast and vice versa were service lines of the big companies.

From the point of view of the cable companies the tariff question was interesting in that, despite being members of the union, they themselves were the ones who decided their tariffs, except in those occasional cases where the government granting the license had set an upper limit. The cable companies adopted the union's regulations (*règlements*) and the other appendix, but they did not generally go along with the tariffs.

One weakness in the functioning of the union was undoubtedly the fact that at the union's meetings the companies did not have the right to decide on matters that concerned them. They simply had the right to attend and to speak in sectional and plenary sessions. In the decision-making process their interests were represented by their governments. The companies could attempt to influence matters through the debates and discussions, of course, and they could also try to influence official representatives. Bearing in mind the important role that submarine cables played in the world's telegraph network, a backseat role must have been very frustrating for the companies that owned and ran these cables.

Monopolies, Competition, and the Telegraph Users

By granting exclusive operating rights and by allowing subventions, governments effectively caused telegraph monopolies to be established in many parts of the world. It is ironic that just when economic liberalism was taking over as the governing principle in world trade, this important new means of bringing the whole world together—the submarine telegraph cable—should have developed in such complete opposition to the prevailing philosophy. It was not easy for anyone who made use of the intercontinental telegraph to avoid supporting a monopoly.

It was quite natural that those who needed to make frequent use of the telegraph, such as trading houses for example, should have demanded low tariffs almost from the very time that the submarine cables first came into use. The granting of privileges increased these demands because it was easy for outsiders to think that a company operating under the protection of exclusive rights could charge whatever rates it pleased. At least in public there was apparently little understanding of the fact that the cables and other equipment were expensive and that in order to stay in operation the companies had to build up reserve funds, which were indeed unusually important in such a high-risk business. Nor was it recognized that some of the cable companies were financially very insecure businesses, especially those European companies operating in South American waters, for all their exclusive rights.[24] These companies found out for themselves how difficult it was to run a telegraph service in countries where the standard of education was generally modest and, moreover, where political unrest interrupted their business. They also came to realize how inadequate the available telegraph technology was in tropical conditions. In many cases the returns made by these companies fell far short of their expectations; losses forced them to reorganize their capital and always led to financial loss for the owners. On the stock market, shares in these companies were trading at only a fraction of their nominal value.

The cable companies did not, however, have monopolies in all areas, and where there was competition between companies it worked to the advantage of the users. The most important area from the point of view of world trade was the North Atlantic, where there were no exclusive rights because neither Britain nor the United States had made any such concessions. After the Anglo-American cable had been opened between Ireland and Newfoundland in 1866, the same company laid new cables on the same route in the early 1870s. Other new cable connections were made across the Atlantic too. By 1914 there were a total of 17 cables connecting Europe and North America beneath the North Atlantic.[25] As capacity rose, competition also increased, despite the fact that some companies grouped themselves into pools to try to keep it in check. Competition was particularly fierce in the Atlantic in the mid-1880s after the American firm, the Commercial Cable Company, laid a cable between England and the United States in 1884.[26] Before the First World War, competition on the Atlantic route, especially between France and Germany, helped to keep prices down.

The service between Europe and South America was run exclusively by the Brazilian Submarine Company, a British firm, for 20 years. The company secured the necessary rights from Portugal and Brazil in 1872. On the coast of Brazil, the Lisbon to Pernambuco cable joined a cable belonging to another British firm, the Western & Brazilian Company, that ran from Pará on the Amazon delta right down the coast, finishing up on the Uruguay border. South of this border two other British companies ran the lines, taking the service on to Buenos Aires. From here another important landline ran through the Andes to Chile. For the cables that were laid along the coast there were no serious competitors: it is true that the Brazilian government owned some landlines, but service on them was so poor that they posed very little challenge to the submarine cables. All four companies that operated on this coast worked together under a "joint purse" agreement.[27]

At an earlier stage, in 1873–1874, this same consortium, which had been set up by the Western & Brazilian Company, had tried to connect North and South America via the Caribbean. A cable was laid, but it worked for only a short time. The enterprise, undertaken with high hopes, ended in complete failure, and a direct connection between North and South America was not realized. For this reason, all telegraph traffic between North and South America had to travel via Europe.[28]

A direct line between the two Americas was opened only in 1884, when a United States enterprise, the Central and South American Company, laid a cable from the west coast of Mexico via Panama to Lima. There the cable connected to a cable operated by the West Coast of America Company, again British, which ran south to Santiago de Chile, and the connection between the west and east coasts was provided by the landline through the Andes. This meant that traffic between Europe and South America could now also travel via North America. As soon as this route was opened there was at once competition for the South American telegraph traffic, and this resulted in a reduction in the prices charged by the companies for the service.

The monopoly granted by the Brazilian government for the telegraph service to Europe came to an end in 1891. When it did, the government immediately granted another company, the South American Cable Company, the right to lay a cable from Europe to Brazil. The cable ran from Cadiz, Spain, via the coast of Africa in Senegal to Pernambuco. When this cable was opened in 1892, prices again fell.[29] Later, in 1906, the company was taken over by the French government and renamed Compagnie des Câbles Sud-Américains.[30] Before the First World War several other lines were opened for traffic, both to South America and to other parts of the world. In 1891 the French company Société Française des Télégraphes Sous-Marins established a cable from the island of Martinique down the coast of Guyana to Brazil. From Martinique there was also a French cable to New York. This line between New York and Brazil could have been very useful and important, but it was so poor that it offered no competition to alternative routes. In 1911 the German company Deutsch-Südamerikanische Telegraphengesellschaft laid a cable from Bremen to Brazil via Monrovia, for which it received financial support from the German government. Another new line was the one opened by the British-owned Western Telegraph Company, which increased its own capacity considerably when in 1910 it began to operate a new cable connection between Madeira and Buenos Aires via Ascension Island. Traffic carried by this company between Europe and South America was henceforth mostly directed along this line. An agreement signed by the Western Telegraph Company with the government of Argentina in 1909 required a reduction of 15 percent in the prices currently being levied by the company, a move that was extended to cover Uruguay, Paraguay, Peru, and Bolivia.[31]

By 1914 there were three telegraph routes in operation between South America and North America, two cables via Panama and the third via the Caribbean. Between Europe and South America via the South Atlantic there were five cables: three British, one French, and one German. These were all in competition with one another, so that in practice any reduction in prices introduced by one company had to be followed by a similar reduction by all the others. In fact, there would have been even more cable connections available and therefore more competition had governments not at an earlier stage in the development of the service made exclusive concessions that prevented them from issuing any other operating permits. Brazil is an especially clear example of this: the Brazilian government had granted monopolies against which later attempts by both German and American companies to extend their networks foundered.

Those parts of the world that from the European point of view were relatively free of competition were Africa and the Far East. Both these areas had in common the fact that they were largely governed by the British, and certainly in order to reach them from Europe it would have been necessary to touch on lands that were under British control. This alone was sufficient to restrict competition. Africa itself, apart from the southernmost part, was also of little interest to the cable companies because it was relatively insignificant in economic terms. Such cables as were laid to the coast of Africa were heavily subsidized by the government of the country in question.

Between Europe, India, and Australia, the Eastern Telegraph Company, the Indo-European Telegraph Company (a subsidiary of the Eastern Company), and the Eastern Extension Telegraph Company—all British—held a monopoly until 1904. After that a line owned by the British government, the All British Line, which ran from Vancouver to New Zealand and Australia and thence to England via the Cape, offered competition to the Eastern companies at least over some stretches. In 1903 the Commercial Pacific Cable Company laid a cable from San Francisco to China and Japan via Hawaii. This was nominally an American company, but in fact it was half-owned by the British Eastern companies and one-quarter-owned by the Danish Great Northern Telegraph Company: only 25 percent was formally in American hands. At the time, this structure of ownership was a closely guarded secret. The European companies involved in it had founded the company precisely in order to be able to prevent competition for traffic to Asia with the services provided by their own lines to the east.

One of the partners in the monopoly on the Far Eastern service, as mentioned, was the Great Northern Telegraph Company. Cables belonging to this company linked the northern countries on the Baltic to Russia's Siberian line, which began in St. Petersburg and reached as far as Vladivostok. From there it continued to Japan and China, again under Danish ownership and management. The Great Northern had obtained the license to run the Japanese telegraph service from the government of Japan in 1870. No license had ever been granted by the Chinese government, but the cable was simply brought ashore in the foreign quarter of Shanghai in the dead of night. Not until about ten years later were the Chinese officially informed of the existence of the cable.[32] In 1882 the Great Northern was given a 20-year monopoly on its international service out of Japan.[33] From the start, the Great Northern and the British companies had always worked together and to a certain extent operated under a "joint purse" agreement.

The rights and other concessions that the companies had acquired over the years had made their positions so strong that it was impossible for other companies to compete in the eastern service. In practice they held a monopoly position and could determine the prices they would charge.

This situation did, however, provoke some bad feeling, especially among businesses that had trade links with the Far East. Criticism of the cable tariffs began to be heard openly in the mid-1880s, but the criticism became much more vociferous because of the so-called Chefoo telegraph treaty in 1887. This was an attempt by the companies to persuade the Chinese government to grant them new privileges that would have made competition even more impossible. In order to prevent its ever coming into effect, chambers of commerce in both Europe and the Far East bitterly criticized the proposed treaty, and this in turn led to the whole monopolistic structure of the Far Eastern service coming under scrutiny.[34] At the beginning of the 1890s various governments began to be concerned about the conditions governing telegraph transmissions to the Far East. The monopoly was particularly irksome to countries that did not have their own cables, such as France and Germany, the United States, and Japan itself. In 1879 Japan had become a member of the ITU, and the country had therefore begun to have a say in how the transoceanic telegraph was run. China became a member in 1908.

It was the British government, however, that played the most important role. Some chambers of commerce, various merchant groups, and different trading houses brought pressure to bear not only on the telegraph companies but also on the Board of Trade and the Foreign Office to get them to do something about the problem, since it was assumed that the British government did have some influence in the matter. These departments in London passed the problem on to the postmaster general for him to find a solution. Because the British companies had by far the greatest length of cable in the world, the views and actions of the postmaster general came to be of worldwide significance.

In 1889 the Foreign Office for the first time organized a meeting to discuss the Far Eastern telegraph tariffs. The meeting was attended by the British cable companies that operated in the area and by the London Chamber of Commerce. These negotiations marked the beginning of a deliberate attempt in London, using the expertise of the Post Office, to find a compromise between the position of the telegraph companies and the demands of the public who wished to use the telegraph service. What had to be acknowledged was not only that the interests of the public should be taken into consideration but also that the cable companies had rendered valuable services to the empire and that without financial security they could not continue to operate.

From 1889 onward the question of the tariffs for the telegraph service between the Far East and other parts of the world was a recurring issue. Negotiations resulted in no significant concessions from the companies until the early 1900s. Then two new lines came into service: one was a Pacific cable brought into use by the Commercial Pacific Company; the other was a new landline that ran from Irkutsk to Peking across the Gobi desert. The latter line opened immediately after the Boxer Rebellion in China; it had been finished just before the rebellion but had been partly damaged in the uprising. In addition to these lines, the Netherlands and Germany laid their own short cables to the Far East. The new telegraph companies together with the telegraph authorities urged renewed negotiations and some discussion of tariffs. These negotiations took place in the years 1902–1904 and resulted in an agreement for very much lower tariffs.[35]

In 1913 the tariffs for the eastern service were once again the object of negotiations. New cables had been laid and demands for reductions in the tariffs were heard as before. This time the negotiations were led by the British General Post Office, and it was largely due to their negotiating skills that agreement for a further reduction in the tariff was reached. Additionally, in a wide-ranging agreement between the two sides success was achieved in fixing a scale of charges according to which the price would depend on the growth of traffic: the greater the traffic, the lower the price.[36]

The result of a series of negotiations after the middle of the 1880s was thus a steady fall in the level of tariffs to the east. In 1887 the price per word between Europe and China was F 8.50; in 1913, F 4.40. There was a similar fall in the tariff for the service between Europe and Japan.

What part then did the International Union play in all this? In fact it had nothing to do with setting the international tariffs. These were negotiated between the cable companies and the various national telegraph authorities. The only part that the union did play was in various meetings to confirm the outcome of some of the negotiations, as happened in Budapest in 1896 and in Lisbon in 1908.[37]

THE OVERALL CHANGE IN THE CABLE CHARGES, 1876–1914

Table 5.2 shows the tariffs for some of the most important international cable routes between London and other cities in 1876 and 1914. London is appropriate to use as the starting point for this comparison, for it was undoubtedly the center of the worldwide telegraph business before the First World War. The year 1876 has been chosen for the comparison, because the change was generally made to the single-word tariff in that year.

Tariffs between other European countries and the rest of the world were not so different from those out of London that they would have had any very significant effect on business. Originally Europe was counted as just one area for the purposes of the Far Eastern service; similarly, the tariffs between Europe and North or South America were calculated in terms of continents rather than countries.

These figures show clearly how competition, where it existed, had the effect of reducing prices, first in the North Atlantic service and later in the South Atlantic. By contrast, there was a relatively more modest fall in prices between 1876 and 1913 on the route between Europe and the Far East,

Table 5.2
Changes in Tariff Rates per Word, in Francs, between 1876 and 1914

From London to:	1876	1914
New York	5.00	1.25
San Francisco	6.05	1.90
Rio de Janeiro	16.30	2.925
Buenos Aires	19.63	3.125
Valparaiso	25.88	3.45
Bombay	4.70	2.25
Hong Kong	10.00	4.40
Shanghai	10.00	4.40
Nagasaki	10.00	4.88

where there was no competition. Even this modest reduction came about only after severe pressure was brought to bear on the cable companies.

In 1876 sending a telegraph from one continent to another was so expensive that it was out of the question for most private individuals. Even the wealthier members of society would use the international telegraph only in very exceptional circumstances. Comparison of salaries at the time with the price of sending a telegraph brings the point home: in 1876 a good daily wage for a European artisan was around three French francs. The figures show that one day's pay would not have been sufficient to cover the price of even a one-word telegraph.

By the outbreak of the First World War tariffs had come down so much that most working people could, if really necessary, have afforded to send a telegraph from one continent to another. We can thus reasonably say in conclusion that three factors globalized the telegraph business: the technical development of the lines and the cables, the common framework for the operation of the telegraph provided by the ITU, and the decline in the cost of using the service.

Notes

1. One or two examples of this can be given here. The line built by the Prussian government was opened to the public as early as 1849, but it could only be used if there was surplus capacity, the needs of state officials taking priority. Otto Frank, *Das Buch von der Weltpost* (Berlin: Verlag von Herm. J. Meidinger, 1894), 256. In France the telegraph was opened by the government and remained exclusively for government use until 1850. Only then was it opened to the public. *Lois e Règlements, Années 1833 à 1854* (Paris, 1855), Law 29 (November 1850). Outside Europe, in Brazil the first telegraph lines were established in 1852 to make it possible for the emperor to have contact with his military units. Victor M. Berthold, *The History of the Telephone and Telegraph in Brazil, 1851–1921* (New York: American Telephone and Telegraph Company, 1922), 4. In Japan the telegraph was opened in 1869, originally only for government use. Yuzo Takahashi, "The Beginnings of the Telegraph System in Japan," presented at the Colloquium Télécommunications, espaces et innovations au XIXe et XXe siècles (Paris, January 1989), 4–5.

2. The French decree mentioned in note 1 above.

3. *L'Union Télégraphique Internationale (1865–1915)* (Berne: Bureau International des Administrations Télégraphiques, 1915), 3–4.

4. Frank, *Das Buch von der Weltpost*, 257; *L'Union Télégraphique Internationale*, 3–4.

5. *L'Union Télégraphique Internationale*, 4–5; George Arthur Codding, Jr., *The International Telecommunication Union: An Experiment in International Cooperation* (Leiden: E. J. Brill, 1952), 14.

6. *L'Union Télégraphique Internationale*, 8.

7. Codding, *The International Telecommunication Union*, 18–19.

8 A. J. H. de Clerq, *Recueil des traités de la France*, vol. 9 (Paris, 1880), 83–107, 235–257; Codding, *The International Telecommunication Union*, 20–23.

9. *Documents de la Conférence Télégraphique Internationale de Vienne* (Berne: Union Télégraphique Internationale, 1868).

10. *Documents de la Conférence Télégraphique Internationale de Rome* (Berne: Union Télégraphique Internationale, 1872).

11. Jorma Ahvenainen, *The History of the Caribbean Telegraphs before the First World War* (Helsinki: Suomalinen Tiedeakatemia, 1996), 9–11.

12. Jorma Ahvenainen, *The Far Eastern Telegraphs: The History of Telegraphic Communications between the Far East, Europe and America before the First World War* (Helsinki: Suomalinen Tiedeakatemia, 1981): 41.

13. For example, Ahvenainen, *The History of the Caribbean Telegraphs*, 12–22.

14. Robert Boyce, "Submarine Cables as a Factor in Britain's Ascendancy as a World Power, 1850–1914," in *Kommunikationsrevolutionen: Die neuen Medien des 16 und 19 Jahrhunderts*, ed. Michael North (Cologne: Böhlau, 1995), 86.

15. Ahvenainen, *The History of the Caribbean Telegraphs*, 12–13.

16. The relevant conference documents used here are in the Danish State Archives, in the Koloniernas Centralstyrelse series. Kolonialkontoret, Gruppesager, Telegraf og Telegrafanlaeg, vol. 3. Denmark joined the concession to Balestrini in 1865 on account of its colonies in the Caribbean.

17. Portuguese concession, November 18, 1872, in Concessions and Agreements, Brazilian Submarine Telegraph Company, Cable and Wireless Archives, Porthcurno, United Kingdom. The concession was given to the Brazilian Baron de Maya, who worked as the company's representative. The Telegraph Construction and Maintenance Company immediately transferred the concession to a subsidiary, the Brazilian Submarine Company, founded for the purpose. Here it is relevant to mention that in international telegraph traffic it was normal at this time to calculate the tariff in French francs on the basis of a 20-word message. This had been decided at the above-mentioned international telegraph conference held in Paris in 1865.

18. House of Commons, May 19, 1879, *Accounts and Papers*, 1878–1879, 42:911; Hugh Barty-King, *Girdle Round the Earth: The Story of Cable and Wireless and Its Predecessors to Mark the Group's Jubilee, 1929–1979* (London: Heinemann, 1979), 69–70.

19. Ahvenainen, *The History of the Caribbean Telegraphs*, 60–65.

20. Convention Télégraphique International de St-Pétersbourg et Règlement et Tarifs y Annexés (Berne, 1875).

21. The number of companies has to be calculated from the records of the union conferences. The names of companies that had joined the union were also announced in the Berne Notification.

22. *Documents de la Conférence Télégraphique Internationale de Londres* (Berne: Union Télégraphique Internationale, 1879), Tableau B.

23. Berne Notification January 30, 1877.

24. These companies were the Western & Brazilian Company, the West Coast of America Company, the London Platino Brazileira, and, in the Caribbean, the West India and Panama Company and the French Compagnie Française des Câbles Sous-Marins.

25. *Journal Télégraphique* 38 (1914): 147–171.

26. *Journal Télégraphique* 8 (1884): 113.

27. Agreement with Brazilian Submarine, Western & Brazilian, London Platino, and the Brazilian and Montevidean Company, 1873–1874, in Concessions and Agreements 1872, Western & Brazilian, Cable and Wireless Archives, Porthcurno, United Kingdom.

28. Ahvenainen, *The History of the Caribbean Telegraphs*, 37–41.

29. Berne Notification, November 1, 1892.

30. Board of Trade, London. Register of Companies No. 34363.

31. *Boletín Oficial* (of Argentina), September 30, 1909.

32 Ahvenainen, *The Far Eastern Telegraphs*, 44, 92–93.

33. Ibid., 67.

34. Ibid., 118.

35. Ibid., 180–185.

36. Ibid., 203–204.

37. *Documents de la Conférence Télégraphique Internationale de Budapest 1896* (Berne: Union Télégraphique Internationale, 1897); *Documents de la Conférence Télégraphique Internationale de Lisbonne 1908* (Berne: Bureau International des Administrations Télégraphiques, 1909).

6

The Origins of Cable and Wireless Limited, 1918–1939: Capitalism, Imperialism, and Technical Change

Robert Boyce

In January 1928 the Imperial Wireless and Cable Conference convened in London to address an imminent crisis threatening communications between the far-flung components of the British Empire. The result was the most radical reorganization of imperial communications since high-speed links were established in the 1860s, with nearly all of Britain's overseas operating companies merged into a single firm. The holding company was called Cables and Wireless Limited, and its operating subsidiary was Imperial and International Communications Limited, later renamed Cable and Wireless Limited. Contemporary American and British observers perceived this development as though in mirrored images. American officials and businessmen regarded it as an aggressive move by Britain intended to consolidate its near monopoly hold on world communications and bottle up American commerce.[1] Their British counterparts regarded it as a step toward imperial unity and a defense against the globalizing ambitions of American multinational enterprise.[2] There was, as will be seen, some substance to both views. But notwithstanding the prominence given to Anglo-American rivalry in recent historical accounts of this period,[3] these are caricatured explanations that should not be allowed to obscure the central influences on the decision to create Cable and Wireless.

One such influence was technological advances, particularly in the field of radio telegraphy in the 1920s, which profoundly affected the profitability of communications systems relying on the older submarine cable technology. A second influence was the ideological preference among British decision makers for commercial or capitalist solutions, which deterred them from seriously contemplating a different approach to the problems brought on by new developments in radio telegraphy. A third influence was the growing *dis*-unity within the British Empire, brought on by the increasing reluctance of the self-governing dominions to collaborate with Britain except on a commercial basis. Although on the face of it the merger was an imperial project, in fact it was essentially a British initiative. And rather than being inspired by a broad imperial vision, it was in fact intended to safeguard Britain's strategic defense and security, its cable and radio telegraph companies, and the London financial institutions that largely owned the communications companies and relied on their services. As it happened, creating the merger was relatively straightforward. Satisfying the divergent interests, however, proved to be an altogether harder task.

Background to the Merger

In 1918 British submarine cable companies dominated most of the world's intercontinental communication routes, with one important exception. This was the highly lucrative North Atlantic route, linking Britain, the United States, and Canada. The route had been pioneered by an Anglo-American consortium in 1866 and was dominated by the British-owned Anglo-American Telegraph Company, predecessor of the Eastern and Associated Telegraph Companies, until the turn of the century, when it abruptly passed into American hands. Anglo-American had been obliged to give way because of its inability to generate business in the American market. There, two firms, the Western Union Telegraph Company and the Mackay-Bennett system, largely controlled the internal telegraph network[4] and decided to aggressively expand into international communications, exploiting their monopoly within the United States to drive out competition on their international routes.[5] All outgoing messages from the United States were directed to their own Atlantic cables, while all incoming messages sent on British cables were charged penal rates to be delivered within the United States; Western Union simply refused to collect messages for competitor cables on the Atlantic route.[6] The Anglo-American Company had no effective means of responding, since the United States, unlike Britain, refused to sign the International Telegraph Convention, which outlawed such discrimination.

The British authorities, aware of the vital importance of British cable companies to the country's commercial and imperial strength, supported their efforts to expand globally and took an indulgent view of their practice of colluding with competitors in rate-fixing agreements in order to maintain profits. But the British General Post Office (GPO), which had owned and operated the domestic telegraph system since 1870, sought to ensure that as many communications lines passed through London as possible and hence generally welcomed the landing of foreign cables in Britain.[7] For the same reason, when collecting or delivering messages, it did not discriminate between British and foreign telegraph services, except in cases where British senders of telegrams to North America did not specify a preferred service.[8] Thus as late as 1910 five of the thirteen commercial cables on the North

Atlantic route were British owned and eight American owned, the other four cables being owned and operated by the German and French national post and telegraph systems. But in January 1911 the two British companies operating on the route, the Anglo-American Telegraph Company with four cables, and the Direct U.S. Cable Company with one cable, amalgamated with Western Union (with whom they had hitherto operated a cartel), as the result of which all thirteen commercial cables became American owned or leased.[9]

In July 1868 the GPO had entered into an agreement with Anglo-American whereby it would not ask customers which cable they wished their telegrams to be sent on and would hand over to the company all "unordered" telegrams to North America (including Canada). This agreement, renewed in June 1901, was due to continue until 1920.[10] Even after the Anglo-American Telegraph Company passed into American hands, it was maintained and extended to 1928, to the dismay of British imperialists.[11] Nonetheless, American commercial interests continued to represent the British and continental European telegraph systems as hostile state-owned monopolies and to contrast their operations unfavorably with the American telegraph industry, which was supposedly governed by free market principles.[12]

The First World War increased American dissatisfaction with European "monopolists," when at the start of hostilities Britain cut the German cables in the North Atlantic and proceeded to divide them up with its ally, France.[13] This cut off Germany from direct communication not only with North and South America but also with much of the Mediterranean and Africa. It also cut off the United States from direct high-speed communications with Germany and the other Central Powers. The result was that for the duration of the war messages in either direction had to pass through British- or French-controlled cables and were thus exposed to interception, censorship, and delay, or be sent via radio telegraph and similarly face interception by the British and French. After the war, therefore, the disposal of the ex-German cables became a source of bitter dispute between the United States and its erstwhile allies, with Britain and France insisting upon treating them as legitimate spoils of war and the United States demanding their restitution to Germany.[14]

Meanwhile U.S. Navy officers, along with State Department officials, urgently demanded a national commitment to construct a comprehensive network of international communications to liberate the United States from British domination.[15] President Wilson, who fully shared their suspicions of Britain and frustration with America's subordinate position, supported their demand while in Paris to attend the peace conference.[16] The result was RCA, created in the autumn of 1919, with Admiral William Bullard, the superintendent of the Navy radio service, appointed an official observer with a nonvoting seat on the board.[17] The practical results, however, proved disappointing. RCA erected several high-powered radio telegraph stations for overseas business and soon acquired a large share of the United States–East Asia traffic. But after Admiral Bullard's retirement in 1922, the Navy's influence over company policy dwindled,[18] and, once the postwar slump ended, the company turned away from international communications in order to concentrate on more lucrative opportunities within the American domestic market.[19]

The American firm with the largest stake in international communications at this time was the Mackay group of companies, which operated several North Atlantic cables through its subsidiaries the Postal Telegraph Cable and the Commercial Cable companies, as well as the cable linking the United States with Hawaii, the Philippines, and East Asia through its subsidiary the Pacific Commercial Cable Company. But Washington's confidence in the Mackay group had been shaken in 1920, when a congressional enquiry learned that the Pacific Commercial cable, constructed in the aftermath of the Spanish-American War in order to liberate the United States from dependence on British cables, was from the beginning effectively controlled by the Eastern and Associated Telegraph Companies, a British consortium.[20] Another potential flag carrier had by this time appeared when International Telephone and Telegraph (ITT) was created in 1920. Having begun as a small service provider in the Caribbean basin, it forged an alliance with AT&T in 1925 wherein AT&T devoted itself to domestic communications while ITT agreed to concentrate on overseas markets.[21] Under the dynamic leadership of Sosthenes Behn, ITT rapidly expanded through the acquisition of foreign equipment manufacturers, telephone and telegraph operating companies, and several international cable companies.[22] In February 1926 it acquired All America Cables, a major service provider linking the United States with Central and South America. In March 1928 it agreed to a merger with the Mackay group, thereby becoming a major competitor on the North Atlantic and Pacific routes.[23] Always aggressive, it imitated the American firms that had captured the North Atlantic route before the war by using its monopolistic position in certain overseas markets, including Spain, Portugal, and Argentina, to freeze out foreign competitors on its international routes.[24] But despite ITT's remarkable overseas expansion, it failed to gain the trust of Washington officials because its takeover activity regularly involved the coopting of foreign nationals onto its main board or those of its subsidiaries.[25]

Behn defended his actions, claiming that they were necessary to overcome foreign opposition to his often politically sensitive acquisitions.[26] This was probably a fair assessment, but it was also true that ITT and other American companies owed their first loyalty to shareholders, whose interests did not always coincide with Washington's policies. A similar criticism could be made of British communications companies. The Eastern group of companies, including the Western Telegraph Company, which operated along the east coast of Latin America, and the Eastern Extension, Australasian and China Telegraph Company, which dominated routes from India eastward, constantly represented themselves as loyal servants of the British Empire. The close support they received over the years from the Royal Navy and the General Post Office seemed ample proof of their assertions. Nevertheless they had long employed predatory pricing to deter competition or force competitors into joint purse arrangements in order to secure monopoly profits. The indifference of the Eastern Extension company to cable users in Britain's own antipodean colonies after its cable reached them in 1872 eventually led to demands for a publicly owned alternative service.

In 1879 Sir Sandford Fleming, the chief engineer and surveyor of the first Canadian transcontinental railway, entered the scene with a proposal to construct a Pacific cable that would give the antipodean users an alternative route to Britain and the rest of the world via Canada and the North Atlantic cables. The Eastern Telegraph Company, with tacit support from the British government, sabotaged Fleming's repeated attempts to secure the necessary public backing for his scheme. Eventually, in the troubled international conditions that accompanied the Second Boer War of 1899–1902, the British government abandoned its opposition to a transpacific cable. The Pacific Cable Company, jointly owned by Britain, Canada, Australia, and New Zealand, was created and its cable began service in 1902.[27] The first public utility of the British Empire, it was also the only one and demonstrated the severe limits of imperial cooperation.[28]

But before this became clear, the First World War reconfirmed the view of British military and naval authorities that the cable network, providing secure communications with the dominions and colonies, was essential for Britain's survival.[29] The war favored the cable companies by stimulating a large increase in traffic, which remained substantially greater after the armistice. They also received firm backing from the British government, which was prepared to antagonize Washington by lending diplomatic support to the Eastern Telegraph Company's efforts to retain control of South American business and preferential use of the Azores, thus limiting direct American access to South America and the Mediterranean.[30] Despite the aggressive nationalism of the U.S. Navy, State, and Commerce departments, American competition outside the Atlantic and the Central and South American routes remained limited. And as late as 1925, the future of Britain's commercial cables as well as the state-owned Imperial Atlantic cables and Pacific cable seemed secure. Even the technical changes affecting overseas communications for the time being favored cables as much as radio.

With the first transoceanic radio signal recorded in 1901 and the successful transmission of one-way voice messages from Canada to aerials atop the Eiffel tower in 1914, the outlines of a new revolution in overseas communications were already apparent before the First World War. In 1912 the British government was sufficiently encouraged by the technical progress to announce the construction of eight high-powered radio telegraphy stations capable of communicating with the main outposts of the empire. But at the outbreak of war in 1914 only two publicly owned stations had been constructed, and it would be another ten years before radio telegraphy became a major factor in overseas communications.[31]

In the meantime, important advances continued to be made in cable technology. The most notable development came in 1921 when the continuously loaded cable, used before the war in several German and Scandinavian telephone systems, was adapted for undersea telegraph transmission. While adding only slightly to the cost of manufacture, loaded cables, combined with duplex or multiplex equipment for simultaneous two-way or multichannel operation, provided at least four times the capacity of older cables.[32] The first to apply the technology commercially was Western Union, which ordered a loaded cable from the British firm Telcom in 1923 for a new 2,300-mile line between New York and the Azores. This was followed shortly afterward by a second Western Union loaded cable between North America and England, and another for the Deutsche-Atlantische Telegraphengesellschaft, which continued Western Union's New York–Azores cable on to Emden, near Hamburg, in the North Sea. The Pacific Cable board also took advantage of this technology when it duplicated the longest spans of its cable in 1925, and to good effect. The original 1902 cable had an operating speed of approximately 25 words per minute (wpm) each way, or 50 wpm duplexed; the new cable was capable of transmitting 250 wpm.

The introduction of the higher-speed Western Union cables posed little threat to British commercial companies since they had long ceased to compete on the North Atlantic route.[33] Nor, by and large, did they see a need for heavy investment in technical improvements elsewhere. The Eastern and Associated Telegraph Companies therefore limited themselves to the introduction of the duplexing and "regenerator" systems, the latter using an effective but relatively inexpensive device that enabled signals to be repeated automatically and reduced the need for operators at intermediate stations on the long cable routes.[34] With a monopoly on most of their routes, the Companies were able, through their high rates, to effectively limit demand to the capacity of their existing cables; the income from this service was sufficient to generate high returns for their City-based shareholders.

Thus in 1925 the Eastern Telegraph and Eastern Extension companies together reported gross receipts of £5.5 million ($26.7 million) and net receipts (after expenses) of £2.1 million ($10.2 million),[35] enabling them to pay their shareholders dividends of 10 percent net of taxes while tucking away a similar amount into company reserves. Their sister company, Western Telegraph, despite American competition over part of its routes, fared nearly as well, with gross revenue in 1925 of £1.5 million, net revenue of £0.6 million, and its dividend of 10 percent net unchanged from earlier years.[36] Some idea of the profitability of these firms can be seen from the fact that the Eastern Extension company was originally capitalized in 1869 at slightly over £1.5 million with what was widely assumed to be watered stock. Between then and 1910 it earned its shareholders no less than £6.5 million in dividends, and with only modest additional investment it was still earning its shareholders nearly £1 million a year in 1927.[37]

As late as 1924 little had been done to fulfill the British government's commitment in 1912 to construct an empire-wide radio telegraph network. One scheme for erecting a chain of high-powered longwave stations had been devised by a select committee chaired by Sir Henry Norman in May 1920, but it was rejected by Australia, South Africa, and eventually also by Winston Churchill, the colonial secretary and chairman of the Imperial Communications Committee, as well as by the British parliament.[38] Another scheme based on similar technology was proposed by Sir Laming Worthington-Evans, postmaster general in Stanley Baldwin's first Conservative government in 1923, but it required the cooperation of the Marconi's Wireless Telegraph Company on terms the company found unacceptable. A committee instituted by the first Labour government in January 1924 and chaired by Robert Donald, chairman of the Empire Press Union, put forward a third scheme. Simpler than the previous one, it called for Marconi's Wireless to build four longwave stations for the GPO to use in communicating with the empire, leaving foreign communications for development by the private sector (in practice, Marconi's Wireless). Once again, however, Marconi's Wireless objected to the scheme and, while the question of commercial prerogatives was still under discussion, introduced its new shortwave beam system.[39]

The Committee of Imperial Defence (CID), comprised of senior British cabinet ministers and their chief military advisers, regarded it as a disadvantage that the shortwave system could not communicate simultaneously with different overseas posts, since in the event of war all ports and naval forces would require urgent warning of attack. But for commercial work it promised great economies, since shortwave beam stations were far cheaper than longwave stations to build or run. Within a few months the government decided to proceed with a powerful longwave station at Rugby for defense purposes and to award Marconi's Wireless a contract to build four beam stations, which, if successful, would be owned and operated by the GPO, with Marconi's Wireless receiving 6 percent of the revenue they earned. Construction of the first beam station for service with Canada was soon completed, and, once its counterpart in Canada was ready, services began in October 1926. Service with Australia began in May 1927, and in August of that year the empire system was completed when the remaining two stations began services with India and South Africa.[40]

Since the advent of radio telegraphy, directors of the British cable interests had displayed occasional apprehension at the prospect of competition from the new technology. At the time of the Donald committee enquiry in 1924, for instance, Sir John Denison-Pender, chairman of the Eastern and Associated Telegraph Companies, wrote to Clarence Mackay, Newcomb Carlton, and John L. Merrill, heads of the three principal U.S. cable companies, to propose a global merger or combination in order to confront competition from radio.

> I have been thinking a good deal lately about the position of the Cable interests as regards outside competition and feel that the time has come when something ought to be done. . . . The position, as it appears to me, undoubtedly calls for some measure of combination or working arrangement amongst us all, to enable us to show a united front in competition with wireless[;] instead at the present time we are all drifting further apart and wasting money on schemes of unremunerative competition and intensive canvassing.[41]

Nothing came of his initiative, nor was it ever likely to have done so in view of the close alliance between Newcomb Carlton's Western Union and RCA, and the probable opposition of the U.S. government to such a restraint on trade. But in any case, Denison-Pender appears subsequently to have set aside his apprehension. Claims that radio would supplant cable telegraphy had been heard ever since the turn of the century, but authoritative reports predicted that shortwave transmission, whether "beamed" or not, would remain susceptible to changing atmospheric conditions and hence operational for less than twenty-four hours a day. On this basis, it would be suitable only for nonurgent messages carried at cheaper rates, leaving the cables to carry most of the commercial and news traffic for which full rates could be charged as well as serving the needs of diplomacy and defense. Despite the rapid developments in radio telegraphy, the Eastern and Associated Telegraph Companies did little to improve their own technology. In 1925 they hired a scientific consultant at £500 a year to advise them on the question of introducing "loaded" cables, and in February 1928 they put another scientific consultant on retainer to keep them informed of developments in radio telegraphy.[42] The one substantive step they took came in January 1926 when they agreed to join the All-American Telegraph Companies and the Telegraph Construction and Maintenance Company in forming a private company to undertake research on lines proposed by the last-mentioned firm.[43]

At the annual general meeting of the Eastern Telegraph Company in London in May 1927, Denison-Pender reported net profits of £1.1 million, a figure closely in line with previous years. Referring to the introduction of the beam service, he expressed confidence that it constituted no threat to the cable companies:

> Having carefully watched the development of wireless during the past 30 years, I have no cause to alter the opinion formed and expressed over that period, that cables will remain more reliable than the wireless service. (Cheers.) Gas has not been supplanted by electricity, nor wool by cotton, and the iron and steel industry has not suffered by reason of the introduction of cement. Rather have their respective interests been strengthened by co-operation. The advent of electricity proved a stimulant to the gas industry, and in many cases they now work in the closest co-operation. Cotton and wool interwoven created a new demand throughout the world, and the combination of steel and cement in the form of reinforced concrete has established itself to the benefit of all concerned. The prospect of the return to normal trade will, I feel sure, counterbalance to a very large extent the present-day competition with which we are faced, but the obvious solution in the interests of the public and others is co-operation in some form or another, which would eliminate wasteful and unremunerative expenditure and, at the same time, provide a dual and alternative means of communication. I do not think, however, you need have any misgiving as to the future, for I am confident that, given a fair field for the unrestricted development of the cable industry, it will continue to prove itself the best and most reliable means of telegraphic communication. (Cheers.)[44]

For the balance of the year the revenue generated by the cable companies indicated that they could withstand a good deal of competition from the beam system.[45] Indeed, remarkably, they maintained their dividends while continuing to accumulate reserves well after the date when agreement was reached on merging their cable services with the beam system. The Eastern Telegraph Company reserve fund stood at £4,818,181 in January 1924. By January 1927 it had risen to £6,206,141; a year later, to £6,611,700. By January 1929 it still stood at £6,460,685.[46] Despite competition from both the Post Office beam service to Australia (from May 8, 1927) and RCA, the Eastern Extension, Australasian and China Telegraph Company held up equally well. Its reserve fund rose from £7,727,014 in January 1927 to £8,539,786 by early 1928, and to £9,015,567 by January 1929.[47] In light of these buoyant accounts, it was only very belatedly that Denison-Pender and fellow directors of the cable companies recognized the seriousness of the challenge posed by the beam shortwave system.

Ironically, the one company that faced immediate trouble from the new publicly operated technology was the Pacific Cable Company, owned jointly by the governments of Canada, Australia, New Zealand, and Britain. Unlike its main commercial rival, the Eastern Extension company, which offered inducements to secure the business of important customers in Australia and New Zealand, the Pacific Cable Company, being public, could not discriminate among users of the same category of service.[48] It therefore relied more heavily upon nonurgent traffic in the form of deferrable evening telegrams and weekend letters. This traffic was particularly susceptible to competition from the beam service, and when the latter set its rates for deferred telegrams one-third below those of the cables, the Pacific cable almost immediately lost over 45 percent of its traffic.[49] As it happened, this also occurred at a particularly difficult time for Pacific Cable. Just the previous year, the company's board, over the strenuous objections of its Canadian representative, had chosen to lay a second cable along the main sections of its route. This consumed all the reserves it had accumulated since 1914 and burdened it with a £1.2 million debt, leaving it ill placed to absorb the losses it began to incur after the beam service opened.

Faced with the prospect of the Pacific cable ceasing operation, Stanley Bruce, the Australian prime minister, urgently appealed to the authorities in London to consider the strategic implications of unregulated competition between the cable and the beam service.[50] The Imperial Communications Committee, which reported to the CID, agreed to undertake a general review.[51] In December 1927 it reported its conclusion that nearly all the cables linking imperial territories were required for defense.[52]

Already, however, the British government, at the instigation of Canada, had decided to convene a conference of the four states with a stake in the

Pacific Cable Company. The Pacific cable had been largely a Canadian initiative, pressed upon a reluctant Britain, which saw it mainly as a threat to London's privileged position at the center of world communications.[53] But that was at the turn of the century, when Canadian politics were still strongly informed by the imperial spirit. Now the Canadian government, while professing allegiance to the empire, wanted nothing to do with the CID or the Imperial Communications Committee and wished only to be rid of its interest in the Pacific cable. The recent decision to duplicate the cable, taken against its most express objections, had infuriated the Canadian government, which responded with a public display of anger at Britain and its imperial partners.[54] With the cable threatening to become a charge on the Canadian taxpayer, Ottawa sought a conference in order to persuade the other partner governments to take over Canada's share or dispose of it to British commercial interests.[55] The British government agreed to host the conference, while widening its remit to include the implications of wireless and cable competition for imperial communications as a whole.[56]

The precise details of what transpired in the weeks before the opening of the Imperial Wireless and Cable Conference on January 16, 1928, remain unclear. But it is known that Denison-Pender, at last alive to the challenge from shortwave telegraphy, approached Sir Evelyn Murray, secretary of the GPO, to warn the British government that the cable companies were being driven to the wall by the beam service and might be forced to cease operations and to sell their cable networks to American competitors if the government did not give them protection.[57] This put the government in a predicament. Ministers were aware from the CID's preliminary report that it regarded nearly all the commercial cables to be essential for British defense purposes.[58] They also appreciated that the cable companies' threat could not be casually dismissed. The companies had already written down virtually all their capital investment, had practically no debt, and held more than £20 million in cash reserves, which they could distribute to shareholders before being wound up rather than face a long drawn out struggle with lower-cost radio telegraph competitors.[59] A complete or partial sell-off to American interests, in view of the rapid expansion of American multinationals overseas since 1924, was also bound to seem a real possibility. The government therefore gave Denison-Pender an assurance that the cable companies would be protected.[60] Thus armed, Denison-Pender met Lord Inverforth, chairman of Marconi's Wireless, in December 1927 to propose a merger.[61]

In other circumstances, Inverforth might have brushed aside Denison-Pender's proposal and done nothing until his own firm and the GPO, using Marconi technology, drove the cable companies out of business. But in contrast to the cable companies, Marconi's Wireless was financially weak, having had to write off fully £6 million in investments earlier that year, and indeed having lost £100,000 on the contract to construct the GPO's hugely successful beam stations.[62] On account of this financial crisis, a purge of Marconi's board of directors had taken place, with most of the newcomers being City bankers. They were aware of the low opinion in which the company was held among their colleagues in the Square Mile. They also recognized that they would be extremely unpopular in the City if Marconi's were to side with the state-run GPO, a byword in business circles for public inefficiency, in a life-and-death struggle with the commercial cable companies.[63] But in any case they appreciated that the cable companies held the trump card, now that the government had acknowledged it needed the cables for defense purposes and would not let the companies be closed down or sold off. Thus, as Inverforth pointed out to skeptical shareholders a few years later, Marconi's Wireless had no choice but to negotiate with the cable companies.

> I have seen the statement made in more than one quarter that the advice we gave shareholders was wrong, since if we had declined to merge with the cable companies we had such a powerful weapon in wireless that by developing our services to the fullest possible extent we could have put the cable companies out of business. . . .
> Unfortunately this was not the case. Had it been, we should never have advised our shareholders to enter into the merger. What is forgotten is that when we went before the Imperial Wireless and Cable Conference we were told that the Marconi Company would not be allowed to destroy the business of the cable companies: in other words, that, if we did not come to an agreement, an arrangement would be made between the Government and the cable companies. We should then have found ourselves faced with the competition of the Government and the cable companies acting together, and the cable companies, with their huge reserves, amounting to many millions, could have embarked on a rate war which would have had disastrous consequences to the revenues of our companies in India, South Africa, Canada, and Australia, as well as on the revenue of our own services from this country.[64]

He and Denison-Pender therefore agreed to enter into preliminary talks on a merger and, after the details were worked out, to award the cable companies a majority interest in the postmerger firm. Aware of the embarrassment facing politicians on account of the foundering Pacific Cable Company, however, they also agreed to use their leverage over the government to demand that it hand over the Post Office beam service before they proceeded with the takeover of the Pacific cable.[65]

This confronted the government with another predicament. On the one hand, numerous voices, including those of the Empire Press Union, the British National Committee of the International Chamber of Commerce, the *Economist* magazine, and Sir Charles Bright, son of the pioneer of submarine cables and an authority in his own right, were raised against suppressing competition between cables and radio.[66] On the other hand, rumors circulated of a possible American takeover of the British cable network, a threat assiduously played up by the communications companies and their allies in the press.

Just as the conference opened, Newcomb Carlton, president of Western Union Telegraph Company, and David Sarnoff, vice president of RCA, arrived from the United States for talks at Electra House, headquarters of the Eastern and Associated Telegraph Companies in the City of London.[67] Shortly afterward, Marconi's Wireless submitted to the conference a confidential memorandum with the sensational title, "Extension of American Control over Telephonic and Telegraph Communications throughout the World."[68] Subsequently reports of the proposed ITT takeover of the Mackay Companies received prominent treatment in the London press.[69] In fact, it seems doubtful that any American firm was in a position to purchase the bulk of Britain's undersea cable network. Western Union was financially stretched already, and RCA had little interest in overseas expansion.[70] The only potential buyer was probably ITT, which was able to draw on the financial resources of J. P. Morgan and the National City Bank of New York. Its decision to purchase the Mackay Companies in 1928, however, was almost certainly motivated not by the ambition to expand further abroad, but as a reverse step designed to gain entry into the American domestic market.[71] The Eastern and Associated Telegraph Companies offered no such advantage.

Nonetheless, British observers were sufficiently impressed by the recent acceleration of American foreign direct investment to imagine the worst. Montagu Norman, the governor of the Bank of England, privately complained of American economic imperialism and encouraged British firms to alter their articles of association to forestall foreign takeover.[72] Stanley Baldwin, the prime minister, shared the dislike of American economic expansion then widespread in Conservative circles and, influenced by Norman's warnings, "dreaded American control" of British industry and markets.[73] During the conference, Sir John Gilmour, the chairman, warned British cabinet colleagues that in the event of a failure to secure the merger of Marconi's Wireless and the Eastern Telegraph interests, there was "a serious risk" that all the cables would fall into American hands.[74]

Dominions statesmen were similarly impressed by the seemingly unstoppable U.S. multinational expansion and concerned about becoming dependent on American commercial interests for sensitive defense and security communications.[75] Sir Evelyn Murray, secretary of the GPO, doubted that there was much substance to the rumors that circulated at the time of the conference, but he appreciated their impact. As he later wrote, "The bogey of American competition was very successfully used at the Imperial Conference of 1928 when it was represented that if a wireless merger was not supported by the British government the cables would go into liquidation and would be sold to American interests. There was no doubt that this weighed heavily with the Dominion Governments, though I doubt very much if there was much substance in it."[76]

There was still the question of eliminating competition between cables and radio telegraphy, but this was passed over with relative ease. British ministers proved far more anxious to find a solution to the Pacific cable crisis, which was aggravating imperial relations, and to save the commercial cables from closure or U.S. takeover, than they were to avoid the loss of competition between the services.[77] At the urging of the postmaster general, they initially agreed to a compromise whereby the Post Office retained control of the beam service while pooling revenues with the cable companies in order to ensure that they remained in business.[78] Along with most authorities in the British business community, however, ministers favored a sort of neocapitalism. Taking their lead from Governor Norman of the Bank of England, whose views regularly informed the *Times* business pages, they accepted that with the advent of the large corporation competition was wasteful, since it often left firms incapable of realizing their great potential for economies of scale. Far better, they believed, was the creation of large, stable markets within which firms could justify large-scale capital investment and "rationalize" their operations through mergers and the elimination of redundant capacity.

Thus when Marconi's Wireless and the cable companies affirmed that they must have control of the beam service, ministers soon yielded to their pressure.[79] Sir Campbell Stuart, the Canadian representative at the conference, whose mandate was to dispose of Canada's interest in the loss-making Pacific Cable Company, persistently advocated an "all-in" solution, whereby the commercial cables and the Pacific cable were merged into a single private company along with the Marconi company and the Post Office beam service.[80] Meanwhile, at the conference, Denison-Pender repeated his threat to wind up the Eastern and Associated Telegraph Companies and his demand for control of the beam system.[81] British ministers, seeing no alternative other than to subsidize the cables, reluctantly lent their backing to Stuart's proposal.[82]

Consequences of the Merger

Under the arrangements negotiated at the conference, a holding company, Cables and Wireless Ltd. (commonly referred to in the contemporary press as the Merger company), would acquire the ordinary shares of the Eastern and Associated Telegraph Companies along with the ordinary and preferred shares of Marconi's Wireless and retain all their financial and manufacturing assets including Marconi's patent rights. An operating subsidiary, Imperial and International Communications Limited (I&IC), would acquire the communications assets, including the cables, buildings, radio stations, cable repair ships, and research laboratories. These were to be paid for by issuing shares to the holding company, which in turn would exchange them for the shares of the Eastern and Associated Telegraph Companies and Marconi's Wireless. I&IC would also purchase the Pacific Cable Company for £500,000 and assume its £1.2 million in debts, the West Indies cables and wireless system for £300,000, the GPO's two Atlantic cables for £450,000, and a 25-year lease on the Post Office beam service for an initial payment of £60,000 and annual payments of £250,000. I&IC would create an integrated cable and wireless service for the empire. It would also operate Marconi's nonimperial radio telegraphy service, which was presently to start on shortwave, to New York, Japan, and elsewhere. To ensure that the public interest was properly safeguarded, an overseeing body, the Imperial Communications Advisory Committee (ICAC), comprised of representatives of Britain and its dominions, was to be created. The ICAC would have the right to be consulted on changes in the communication company's policies, the power to refuse rate increases, and the right to reduce rates in the event that the company earned more than its "standard net revenue," which was set at £1,865,000 or approximately 6 percent of its £30 million capitalization. In addition, the British government would retain the right to appoint I&IC's chairman and one other director from names proposed by the cable companies.[83]

The scheme, presented as a new form of imperial public utility, was unanimously accepted by conference delegates and soon approved by the British parliament. With some hesitation, the other interested governments of the empire also approved it, agreeing to cooperate with I&IC and not to license rival systems. Cables and Wireless and I&IC Ltd. came into being on April 8, 1929, and I&IC began operations on September 30, 1929. On the face of it, this was an elegant solution to an extraordinarily complicated situation involving five states and as many operating companies, each with its own constituents or shareholders and individual interests. Yet to call it "a remarkable monument of Imperial cooperation" or "a bold and statesmanlike solution . . . admirably designed to secure the two chief advantages which should be enjoyed by a joint Imperial utility undertaking . . . the economy of private management and the security of public control," as the *Times* described the outcome, was, to say the least, an exaggeration.[84]

If the primary purpose of the conference had been to resolve the intra-imperial dispute over the fate of the Pacific cable, the principal purpose of the merger was to save the cables linking Britain with the distant parts of its empire, which had been rendered temporarily unprofitable and in imminent danger of being closed down as the result of the introduction of shortwave radio technology. The means were in effect to allow the cable companies to take control of their chief competitor, the Post Office beam service, while limiting the introduction of rival communication systems.

Three categories of beneficiary can be identified. The first was the British, Australian, New Zealand, and Indian governments, whose defense and security requirements were served by the maintenance of the cables.[85] The second was the financial, commercial, and mercantile institutions of the City of London, which continued to benefit from an imperial communications network centered on London. For them, rapid access to news of all sorts and above all to market-sensitive information was crucially important; it was very much to their advantage that the imperial system remained centered on London and neither be run down nor subsumed into an American-dominated network. The third category of beneficiaries was the shareholders of the cable companies. These were—once again—largely City-based institutions such as insurance companies, pension funds, and investment trusts, which had been facing the stark alternative of seeing the cable companies wound up or suffer a slow death at the hands of the cheaper radio telegraphy service. Some indication of the gains they stood to make was provided by the huge rise in share prices that occurred once it became apparent that the cable companies would form the core of the new combination. Eastern Telegraph Company ordinary shares, which were as high as 191 pence in 1926, had fallen to 136 pence by the time the conference opened in January 1928 but thereafter rose to 242 pence on reports of a commercial fusion. Interestingly, shares in Marconi's Wireless rose by a similar amount in the spring of 1928, reflecting the expectation that the additional profits from ending the rate war would be equally shared between the cable and wireless companies.[86]

There were also, however, a number of losers from the merger arrangements. One category was the foreign communication companies that might otherwise have been expected to develop services within the British Empire or between the empire and foreign countries and that were effectively barred from doing so by the commitment of interested governments to support I&IC. After its institution in 1929, I&IC began negotiations with Western Union to purchase control of all external communications with the British West Indies. In the spring of 1930 it sought to block proposals from two American firms to provide telegraph and telephone circuits to the Bahamas.[87] In April 1930, I&IC through the ICAC requested Australia to forgo a proposed direct wireless link with the United States. As an alternative, it proposed the establishment of an Australia-Canada link.[88] In October 1930 I&IC through the ICAC called on the postmaster of Hong Kong not to proceed with a wireless service to Formosa, since this would create "needless" competition with its cable link to Japan.[89] In July 1931 it requested the ICAC to call on the Indian authorities to desist from plans to introduce a direct wireless telegraph service with Japan, which would undercut the existing Indo-Japanese cable service. On this occasion the representative of the government of India on the ICAC pointed out that the Indian Radio Telegraph Company, which proposed to introduce the service to Japan, was already two-thirds owned by I&IC. But as the chairman of I&IC acknowledged, the board of Indian Radio viewed the question from an Indian standpoint. They could be more easily dissuaded from proceeding if the ICAC, claiming to speak on behalf of the British Empire, raised objections.[90] ICAC agreed to pressure Indian Radio, but it proved to be too late: Indian Radio had already informed its shareholders of its intentions to establish the new service and was unwilling to face the embarrassment of reversing its decision.[91]

In October 1931 R. H. Brand, a prominent City banker and director of the Canadian Marconi Company, called on the ICAC to block RCA's plans to establish direct wireless telegraph services to India, South Africa, and Australia on the grounds that, if allowed to go ahead, they would undermine the imperial system. The motive for his initiative was, however, to uphold the interests of Canadian Marconi, which had a prior agreement with RCA to provide a route for its communications to and within the British Empire.[92] Already, Canadian Marconi had refused to cooperate with I&IC, claiming that it could not provide adequate service and demanding to resume its connection with the GPO's powerful longwave stations.[93] Despite these difficulties with Canadian Marconi, I&IC intervened through the ICAC to block RCA's expansion into the empire.[94]

In subsequent developments, the company obtained the ICAC's support in 1932 to keep ITT as well as French and Italian companies from obtaining concessions for cable and radio links with Egypt.[95] Later the same year the company secured a concession from the Chinese Nationalist government to construct a beam station in Shanghai for direct communications with Britain. Since 1928 I&IC had held up agreement on wireless telegraphy services with China in order to maintain traffic on its cables; meanwhile, RCA and ITT had expanded into the market. I&IC acted only when it became evident that China would otherwise turn to ITT for telegraph communications between Shanghai and London.[96]

A second category of losers was the foreign countries that would have gained radio links with countries of the empire but for the decision to safeguard the routes operated by I&IC. Thus, in addition to the instances already mentioned, the ICAC raised objections to a radio telegraph link between India and China, another between Australia and Japan, and a third between Canada and East Asia because they seemed likely to divert traffic away from the company.[97]

A third category of losers was the telegraph users in Britain, India, Australia, and other parts of the British Empire. The public overseeing body, the ICAC, intended to see that rates were reduced wherever possible and not to wait until competition forced them down, unlike the commercial cable companies that had been absorbed into I&IC. A start was made in December 1930, when I&IC unilaterally reduced rates between Britain and New Zealand from 1s 11d to 1s 8d a word for full-rate messages.[98] But under the terms agreed at the time the merger occurred, the committee's powers were wholly inadequate for the defense of customers' interests. The most it could do was to decide on rate rises or decreases by an aggregate amount equivalent to one-half the increase in the company's earnings above its "standard net revenue." It had no authority over the running of the company and could not insist that it should invest in technical improvements or reduce its overheads by increasing the efficiency of its administration or operations. Nor could it do anything about the fact that the calculation of the "standard net revenue" was almost certainly based on a large overvaluation of the company's cable assets. Thus the empire's telegraph users were required to subsidize the cable company shareholders and shoulder the main burden of keeping the empire's cables operational. This was scarcely an equitable arrangement, since the users were not the only beneficiaries of the enhancement to British or imperial defense and security. Moreover, while the great majority of the shareholders were City-based financial institutions, it was users in the colonies and dominions who relied most heavily on the company's services. By keeping up the rates well beyond where they would have been had radio telegraphy been allowed to develop freely, the creation of I&IC resulted in a substantial transfer of wealth from the empire's periphery to the center.[99] Dominion and colonial governments appreciated this danger from the start. If they agreed to the creation of I&IC, it was because for Canada this was the only means of ending its commitment to the loss-making Pacific cable, and for Australia, South Africa, India, and other constituents of the British Empire it was the only means of maintaining cables essential to their national security. The moment had passed, however, when they might have supported such an initiative from a spirit of imperial solidarity. For too long they had experienced the effect of monopoly charging by the British cable companies to imagine that a new British combine, even under public supervision, would treat their interests equitably. Accordingly, while cooperating with I&IC in services with Britain, they resisted all attempts by the company to extend its control over the rest of their domestic and external communications. Thus, for example, in April 1930 the company sought but failed to take over the running of the Bahamas wireless telegraphy service. Despite the ICAC's appeals for imperial solidarity, the colonial authorities, reflecting the wishes of local users, firmly refused the company's offer.[100]

The company was scarcely more successful in India, where in 1932 it formed, with the India Radio Telegraph Company, a new subsidiary—the Indian Radio and Cable Company—to operate all the radio and wireless telegraph services to other parts of the empire. The India Radio Telegraph Company nevertheless insisted upon retaining the right to develop its own services to countries outside the empire.[101]

By then, all efforts by the company to take control of Canadian and Australian telegraph services had also failed. Directors of the company hoped to purchase the Canadian Marconi Company, the landlines of the Canadian National and Canadian Pacific Railways, and the telegraph channels of the proposed new AT&T Atlantic cable, combining them with the company's assets into a "Canadian External Communications Company." Their aim was a monopoly of Canada's domestic and external telegraph services.[102] But the Canadian Marconi Company, a private company, partly owned by RCA, persistently refused to be taken over.[103] Officers of I&IC and the ICAC believed that they saw RCA's hand in this obstruction.[104] But if so, it is evident that the federal government in Ottawa was also unenthusiastic about a takeover that threatened to return control of Canada's overseas communications to the British cable interests.[105]

A roughly similar situation arose in Australia. There, a provisional agreement was reached in 1929 between the company and Amalgamated Wireless (Australia) Ltd., which had recently been set up with government support to take charge of all external cable and wireless services. But late in 1930 the government rejected the agreement and announced that henceforth it intended to retain control of external communications.[106]

Unfortunately for I&IC, even before it took over from the beam and cable operators, demand for high-speed communications began to decline, and once the world economic depression got under the way in the winter of 1929–1930, demand slumped further. In the first six months of 1929 estimated revenue from the company's communications services declined over 4 percent from the same period the previous year.[107] In 1930 message receipts declined 13.5 percent from 1929.[108] The company, facing the prospect of having to suspend payment of dividends, sought economies through large-scale redundancies and voluntary wage restraint. Furthermore, at various times over the next few years the company requested that the ICAC agree to reduce the beam rental, to permit rate increases, to abandon cables that were no longer commercially viable or alternatively to keep them open through subsidization, and to take control of the GPO's imperial radio telephony service, arguing that it was inconsistent with the 1928 merger agreement to run it separately from other imperial services.[109]

The company's financial difficulties strengthened the ICAC's hand. It passed on the request to mothball or abandon cables to the Imperial Communications Committee of the CID and meanwhile went well beyond its original remit by insisting upon a wide-ranging review of the company's management and operations.[110] In November 1931 a three-man committee chaired by Wilfrid Greene, a high court judge, submitted its confidential report, which was strongly critical of the company's management. Among other things, it claimed that the board of directors was unreasonably large and excessively remunerated, that the company was substantially over capitalized, and that it had so far failed to integrate the wireless and cable services. The committee considered the possibility of transforming the company into a jointly owned imperial public utility but concluded that the dominions would rather deal with a commercial company than accept any limitation on their national sovereignty.[111]

Armed with the Greene report, the ICAC prodded the company to introduce efficiencies and become more responsive to users' needs. Among other things, this resulted in a reduction in the number of directors from 22 to 14, along with their aggregate remuneration, the appointment of a new chairman, and some reduction of influence of the cable interests.[112] In an effort to expand traffic, the company hired a "publicity manager" in 1934, and in May of that year the company's name was changed from Imperial and International Communications to the shorter and more straightforward Cable and Wireless.[113]

In its 1936 annual report, the ICAC again raised the question of taking the company into public ownership but concluded that this would be unacceptable to the dominions, which feared the loss of their recently won independence from Britain.[114] But in 1937, with the company still struggling to make a profit, dominions representatives met in London and agreed on the introduction of a flat-rate scheme for the empire, which simplified the rate structure while reducing average rates by approximately 10 percent. In return, they accepted the ICAC's proposal of a reduction in the company's "standard net revenue" to increase its chances of earning enough to pay a dividend to ordinary shareholders.[115]

In 1938 the operating company, Cable and Wireless Ltd., finally succeeded in its original ambition of acquiring outright the Post Office beam service. The GPO, however, held out against handing over the imperial telephone service. And in securing the beam service, the company paid over 2.6 million ordinary shares, which gave the British government a direct say in its operations.[116] The ICAC also rejected the company's repeated requests for a subsidy to keep up the strategically essential cables, which the company estimated to cost £450,000—more than its net revenue in 1931, 1932, or 1933 and two-thirds of its net revenue in 1934.[117] The most the committee would do in this respect was to allow a small number of secondary cables to be taken out of service.[118]

The introduction of the flat-rate scheme on April 25, 1937, constituted a significant step toward the goal of cheaper imperial communications. Henceforth, messages were sent for 1s 3d (or 30 cents) a word for full-rate traffic, with a code rate of 10d per word and commensurate reductions for cheaper classes of traffic. This compared with 2s 0d for full-rate messages from London to Australia at the time Cable and Wireless was created in 1928, or 1s 8d to South Africa and 1s 5d to India. On the basis of ICAC estimates, the new rates would save users in Britain and overseas nearly £500,000 a year.[119]

These savings were important, but they were long in coming. And since neither the British nor the dominion or colonial governments were prepared to argue the case for higher taxes to pay for the strategic cables, most of the cost had to be absorbed by the company, which in turn sought to pass on the cost to its customers. This it did partly by severely limiting investment in radio technology. On the eve of the world depression, I&IC estimated non-defense spending on radio research by all British firms or institutions at slightly over £130,000, or $632,000. In the United States, by contrast, the current annual research budget of AT&T alone was $9 million.[120] While the worldwide depression struck the United States more severely than Britain and endured rather longer, the fact that I&IC, with its capital largely tied up in loss-making cables, was responsible for the bulk of British radio research may well have resulted in an equally severe decline in spending. As a result, during the 1930s Britain fell behind other great powers, including the United States, its chief commercial rival, and Germany, its chief potential enemy.[121]

Another consequence was continuing friction with the outlying countries of the empire. The company, backed by the British government, constantly pressured the dominions and colonies to maintain their support by forgoing direct links with foreign countries and instead relying on the company's network.[122] Australia and Canada, the largest, most mature of the self-governing dominions, grew more restive at the sacrifices they were expected to make. Increasingly they opened their communications markets, and, had the threat of war not emerged in the late 1930s, it seems certain that they would have broken altogether with the arrangements worked out in 1928.[123]

The creation of Cable and Wireless vividly illustrates the difficulties that Britain faced in trying to balance the interests of national security, the empire and business at a time of fast-developing technology. The usual justification for the solution arrived at was that it kept in being the vast network of submarine cables, which again proved their worth when war broke out in 1939. Their value at this time can scarcely be gainsaid. There was, however, nothing inevitable about the structure of Cable and Wireless or the way it financed its operations. The actual form it took reflected a balance of influence that favored Britain and the City of London over the dominions and colonies and their telegraph users, and rather than uniting the British Empire, it helped to sustain a constant underlying tension among its constituents.

Cable and Wireless contributed to the world trend away from globalization in the late 1920s and toward imperial or regional blocs. It also adversely affected Anglo-American relations. Confronted with intense competition from the government's beam service in 1927, British cable companies deliberately played up the threat of an American takeover in order to secure government assistance. The threat was almost certainly exaggerated, as indicated above, but nevertheless it seemed plausible and contributed to British and empire-wide support for the Cable and Wireless merger. The specter of a wholesale American takeover of the cable companies in turn contributed to Britain's abandonment of economic internationalism in favor of a strategy of imperial protectionism between 1929 and 1932.[124]

A similar situation arose in the United States in 1929, when ITT and RCA sought to merge and pleaded to be exempted from the 1927 law barring mergers of American cable and wireless telegraphy companies in order to stand up to the British imperial communications giant.[125] The Senate Interstate Commerce Committee began hearings in November 1929. But

Newcomb Carlton of Western Union now protested against the creation of a radio telegraphy monopoly, despite having been himself involved in recent merger talks with RCA. This strengthened opposition in Congress, which, after the Wall Street crash brought to light evidence of widespread corporate corruption, was suspicious of market rigging by large corporations. It therefore resisted pressure to allow the amalgamation.[126] The Cable and Wireless merger, however, renewed the specter of British imperialism and sustained exaggerated notions in the United States of the British Empire's unity and strength, which played into the hands of the large corporations.[127] In 1941, with Britain financially exhausted and in danger of capitulating to Germany, President Roosevelt belatedly set aside his own suspicions of British imperialism in order to establish an informal Anglo-American alliance. Significantly, however, he made it a condition that wherever U.S. forces were stationed in the British Empire, they should be permitted to establish their own communications services.[128]

The American reaction highlights the paradox at the heart of the Cable and Wireless project. On the one hand, it came about largely on account of British fears for the coherence of its imperial defense and security and anxiety over American economic imperialism. On the other hand, it did little to unite the empire or strengthen it in the long run. If anything, like the 1932 Imperial Economic Conference in Ottawa, where British efforts to strengthen imperial unity through internal trade liberalization exposed divergent nationalist trends, the Cable and Wireless project only served to illustrate the difficulty of holding the empire together. Just as the dominions and India wanted access to British markets, so they wanted the benefit of imperial defense, but not at the cost of their independence. Hence they refused to make the concessions demanded from them for an effectively integrated system. By the same token, while British architects of the project insisted that it served the empire as a whole, the practical benefit of their handiwork went disproportionately to London-based British institutions. And while the British represented it as a purely defensive measure, American observers regarded it as an offensive initiative, which increased their resentment of British imperialism and their determination to counter it when the opportunity arose. In the short run, therefore, the Cable and Wireless project served its purpose, but in its broader imperial objective it could be no more than a temporary expedient on account of the costs it imposed upon users, its drag on technical innovation, and the opposition it generated in the United States and the British Empire itself.

Notes

1. See for instance S. C. Hooper to K. A. Bickel, August 8, 1930, Department of the Navy, Reports of the General Board, box 15, Office of the Secretary, Confidential Correspondence, 1927–39 (SC) A6/QH Pacific, RG 80.7.3, National Archives and Records Administration, Washington, D.C. (hereafter NARA); and Leslie Bennett Tribolet, *The International Aspects of Electrical Communications in the Pacific Area* (Baltimore: Johns Hopkins University Press, 1928), which appeared just as the Imperial Wireless and Cable Conference closed. Without commenting on the merger, Tribolet, an American participant at the International Radiotelegraph Conference in Washington the previous year, presented British overseas communications as deliberately designed "to restrain the American eagle. . . . In all phases of their activities the British cable and radio interests are backed by their own government. Private communication companies are often urged to secure monopolistic and other special privileges in foreign countries where privileges would not be granted to foreign companies" (2, 5).

2. "Cable v. Wireless. An Imperial Problem," *Daily Telegraph*, December 30, 1927; Sir John Gilmour, secretary of state for Scotland, to Imperial Wireless and Cables Conference, January 16, 1928, Cabinet Papers, CAB 32/65, I.W.C.(28), first meeting, National Archives, Kew (hereafter NA).

3. See for instance Carl Parrini, *Heir to Empire: United States Economic Diplomacy, 1916–1923* (Pittsburgh: University of Pittsburgh Press, 1969); Frank Costigliola, *Awkward Dominion: American Political, Economic and Cultural Relations with Europe, 1919–33* (Ithaca: Cornell University Press, 1984); and B. J. C. McKercher, ed., *Anglo-American Relations in the 1920s: The Struggle for Supremacy* (Basingstoke: Macmillan, 1990). Michael Hogan, *Informal Entente: The Private Structure of Cooperation in Anglo-American Diplomacy, 1918–1928* (Columbia: University of Missouri Press, 1977), also stresses the rivalry but argues that the two powers eventually reached accommodation in the most sensitive fields, including communications, by the later 1920s.

4. Western Union, which owned four cables on the North Atlantic route, was closely associated with AT&T and the Bell Telephone Company, which together controlled 10 million miles of landlines in North America. The Mackay-Bennett system, which also owned four cables on the same route, was linked to the somewhat less extensive Postal Telegraph (Canadian Pacific Railway) landlines. "Atlantic Cable Arrangements," *Manitoba Free Press*, January 7, 1911.

5. Peter J. Hugill ascribes the U.S. takeover simply to increased American investment in international communications, but this had never affected the British interest in Atlantic route before. Hugill, *Global Communications since 1844: Geopolitics and Technology* (Baltimore: Johns Hopkins University Press, 1999), 43.

6. One reason the British authorities balked at requests from Canada and Australia to join in constructing a state-owned North Atlantic cable was that it would have had no chance of securing business in the United States. A second reason was that it would have reduced the profitability of the British commercial cables. "Cable Rates and Government Cables," memorandum by Sir Henry Babington Smith, GPO, for use of the Cabinet Committee, November 30, 1908, Colonial Office Papers, CO 42/936/1698, NA. Sir

Arthur Balfour, the British foreign secretary, described the process of freezing out British firms to Woodrow Wilson at the Paris Peace Conference. See United States Department of State, *Papers Relating to the Foreign Relations of the United States, 1919*, vol. 4, *Meeting of Council of Ten*, May 2, 1919, 494–495. The policy was frankly acknowledged by the companies. See extract from evidence given by Mr. Willever, Vice-President, Western Union Telegraph Company, before the Board of Railway Commissioners for Canada, July 7, 1920, Records of the Post Office, Post 33/973, M4059/1923, file 1, BT Archives, London; "Inland Facilities for Cablegrams," memorandum handed to the U.S. delegation of the Washington Conference on Electric Communications, October 1920, ibid.

7. F. J. Brown to under-secretary of state, Foreign Office, September 5, 1922, Post 33/1065, M8232/1923, file 51, BT Archives. It was out of concern to see that London was not bypassed that in 1927 the Post Office refused the Western Union Telegraph Company a landing license for a cable from Le Havre to Penzance, for joining to Western Union's transatlantic cable. "Telegraphs from America to France," *Times* (London), September 27, 1927, 16; Sir Robert Donald, "Free Trade in Cables," *Times* (London), September 28, 1927, 10.

8. Committee of Imperial Defence, Imperial Communications Committee, Proposed International Conference at Washington, D.C., I.C.C. No. 288, June 30, 1920, vol. 1247, file 30039, pt. 1, RG 25, Library and Archives of Canada, Ottawa (hereafter LAC).

9. "Atlantic Cable Arrangements," *Manitoba Free Press*, January 7, 1911; Statement in Canada, House of Commons Debates, May 7, 1925, 2962.

10. "Cable Rates and Government Cables," memorandum by Sir Henry Babington Smith, GPO, November 30, 1908, CO 42/936/1698, NA.

11. Memorandum from Department of Post Office, February 11, 1915, vol. 678, file 108009, RG 3, LAC; Minute by Sir John Seely, M.P., December 2, 1909, CO 42/936, file 39,731, NA.

12. See, for instance, Clarence Mackay to Secretary of State, January 28, 1920, State Department Records, Central File, 841.73/39, RG 59, NARA.

13. Aimé de Fleuriau to Lord Grey, March 31, 1916, Committee on Landing Rights, Serial no. 214, Appendix C (3), Post 33/1063, M8232/1923, file bottom, BT Archives; unsigned memorandum, November 3, 1920, France, Ministère des Affaires Étrangères, Série Y Internationale 1918–1940, vol. 621.

14. United States Department of State, *Papers Relating to the Foreign Relations of the United States, 1919*, vol. 4, *Meeting of Council of Ten*, March 6, 1919, 226–227; March 24, 1919, 464–469; May 1, 1919, 486–487; Inter-Allied Committee on Question of International Law Relating to the Capture and Disposal of German Submarine Cables, 1st meeting, March 11, 1919, Post 33/1063, M8232/1923, file Misc. Papers, BT Archives.

15. Hogan, *Informal Entente*, 106–108.

16. Josephine Young Case and Everett Needham Case, *Owen D. Young and American Enterprise: A Biography* (Boston: David R. Godine, 1982), 179, 475–476.

17. Captain S. C. Hooper, member of President's Communications Policy Committee, minute on majority report, December 7, 1933, box 5218, file 811.70/11, RG 80, NARA; Erik Barnouw, *A History of Broadcasting in the United States*, vol. 1, *To 1933* (New York: Oxford University Press, 1966), 57–61; Hogan, *Informal Entente*, 106–118.

18. General Board hearings, "Foreign Radio Policy of the Navy," statement by Commander Hooper, May 31, 1923, M1493, roll 17, RG 80, NARA.

19. This was not explicitly acknowledged but can be deduced from the company's financial returns, see Federal Communications Commission, statement by W. H. Pitkin, vice president and general attorney, Postal Telegraph Cable Company, December 3, 1934, box 5218, file 811.70/31, RG 59, NARA. See also Robert Sobel, *ITT: The Management of Opportunity* (New York: Times Books, 1982), 62.

20. Statement by Clarence Mackay, October 10, 1920, Senate Subcommittee on Interstate Commerce, Cable-Landing Licenses, *A Bill to Prevent the Unauthorized Landing of Submarine Cables in the United States; Hearings on S. 4301*, 66th Cong., 2nd sess. The element of deception can be disputed, but there is no doubt that it prejudiced official attitudes to the Mackay group thereafter. See for instance, H. G. Chilton to Lord Curzon, no. 968, August 18, 1922, Post 33/1065, M8232/1923, file 52, BT Archives; Captain Milton S. Davis, Pacific Communications Officer, San Francisco, to Chief of Naval Operations, Washington, D.C., November 9, 1927, Office of the Secretary, Confidential Correspondence, 1927–39, box 15, file A7-1 (27555), RG 80, NARA; Davis to Chief of Naval Operations, July 21, 1928, ibid., file SC3-6/A6-2 (28283).

21. Sobel, *ITT*, 44.

22. As Daniel R. Headrick puts it, "Behn set out on the greatest shopping spree in history." Headrick, *The Invisible Weapon: Telecommunications and International Politics, 1851–1945* (New York: Oxford University Press, 1991), 202. The acquisitions are listed in Frank A. Southard, *American Industry in Europe* (Boston: Houghton Mifflin, 1931), 43–53.

23. Sobel, *ITT*, 52–53, 63–64.

24. Albert Kammerer to Pierre Laval, no. 41, February 6, 1932, France, Ministère des Affaires Étrangères, Série Y, vol. 619; Papers of Sir Campbell Stuart, vol. 9, ICAC Report to the Governments of the British Empire, appendix 2, Edward Wilshaw to ICAC, September 19, 1930, Cambridge University Library, Commonwealth Society Collection (hereafter Stuart Papers).

25. Testimony of Captain S. C. Hooper, director of naval communications, to Senate Committee on Interstate Commerce, March 15, 1934, box 5218, file 811.70/15, RG 59, NARA; Hooper to Senior Naval Member, Joint Planning Committee, July 23, 1934, box 50, file G.B.419 Op-20-H, RG 80, NARA.

26. Col. S. Behn to Cordell Hull, March 16, 1934, box 5218, file 811.70/15, RG 59, NARA.

27. Robert Boyce, "Imperial Dreams and National Realities: Britain, Canada and the Struggle for a Pacific Telegraph Cable, 1879–1902," *English Historical Review* 115, no. 460 (January 2000): 39–70.

28. Robert Boyce, "Canada and the Pacific Cable Controversy, 1923–1928: Forgotten Source of Imperial Alienation," *Journal of Imperial and Commonwealth History* 26, no. 1 (January 1998): 72–92.

29. Brigadier-General S. H. Wilson, CID, to F. J. Brown, Post Office, enclosing Admiralty paper on "Telegraphic Inter-Communication," April 20, 1921, Post 33/3000, M8943/1930, file 4, BT Archives.

30. Copy of Foreign Office telegram no. 22 to Lisbon, March 22, 1922; Brown to R. A. C. Sperling, Foreign Office, March 23, 1922; Lancelot Carnegie, Lisbon, to Lord Curzon, no. 105, March 25, 1922; Curzon to Carnegie, no. 45, July 1, 1922, Post 33/3000, M4059/1923, file 5, BT Archives.

31. Frances Donaldson, *The Marconi Scandal* (London: Rupert Hart-Davis, 1962), 240.

32. F. J. Brown, *The Cable and Wireless Communications of the World*, 2nd ed. (London: Sir Isaac Pitman & Sons, 1930), 78.

33. In the 1920s Britain again owned and operated two cables on the North Atlantic route, but they were state-owned Imperial cables. One was an ex-German cable that had been cut, diverted, and reconnected in 1914; the other, an old British-owned, American-leased cable, purchased by the government in 1922 to provide a backup for security purposes.

34. "Eastern Telegraph Company: Sir John Denison-Pender's Speech," *Times* (London), June 25, 1924, 21; "The Eastern Telegraph Company, Limited," *Economist*, May 24, 1927, 1139. The technical improvements are usefully summarized in Sir Geoffrey Clarke, "The Future of Telegraphy," *Times* (London), November 23, 1926, 15; "Wireless Telegraph Communication, Internal and External, and Broadcasting, memorandum prepared in Colonial Office, April 1927," Post 33/1441, M15959/1924, file 6, BT Archives; Colonial Office Conference, minutes of fifteenth meeting, May 20, 1927, file 7, ibid.; "The Cable Services," *Chamber of Commerce Journal*, May 4, 1928, 444; and Headrick, *The Invisible Weapon*, 196–200.

35. The sterling/dollar exchange rate between April 1925 and September 1931, while the two countries remained on the gold standard, was £1 = $4.86. From September 21, 1931, to the outbreak of the Second World War the exchange rate fluctuated, the average for the decade being approximately £1 = $4.20.

36. "Cable Companies' Profits," *Economist*, January 22, 1927, 148–149.

37. Commonwealth of Australia, Parliamentary Debates, Session 1904, vol. xxi, 4974–4975, statement by Sen. Staniforth Smith, September 28, 1904. On the allegation of watered stock, see also Sir Sandford Fleming to Rodolphe Lemieux, November 26, 1908, vol. 627, file 65162-4, RG 3, LAC; Lemieux to Sir Herbert Samuel, June 16, 1910, ibid., vol. 678, file 108009-2.

38. "Imperial Wireless Telegraphy," United Kingdom, *Parliamentary Papers, 1920*, Cmd. 777, vol. xviii, pp. 67–97; Post 33/3000, M8943/1930, file PS/28, CID, ICC twenty-fifth meeting, minutes, March 30, 1922, BT Archives.

39. "Empire Wireless," *Times* (London), March 1, 1924, 7; "Empire Wireless: Marconi Company and the Report: Letter to the Prime Minister," ibid., March 8, 1924, 7.

40. Minutes of forty-first meeting of Imperial Communications Committee of the CID, July 29, 1924, Post 33/1441, M15959/1924, file 3, BT Archives; Cabinet Wireless Telegraphy Committee, memorandum by Marchese Marconi, "Historical Survey of the Recent Technical Development of Long Distance Radio Communication" (December 3, 1929), Post 33/5301, pt. 1, M20115/1938, file bottom E, W.T.C. (29)9, BT Archives; W. J. Baker, *A History of the Marconi Company* (London: Methuen, 1970), 204–225.

41. Eastern Telegraph Co. Limited, minutes of the board, February 5, 1924, copy of letter from Denison-Pender to Mackay, Carlton and Merrill, February 5, 1924, Minute Book No. 18, Cable and Wireless Archives, Porthcurno Telegraph Museum, Porthcurno, United Kingdom.

42. Eastern Telegraph Co. Limited, minutes of the board, March 31, 1925; minutes of the joint meeting of the Associated Companies, February 28, 1928, Minute Book No. 18, Cable and Wireless Archives.

43. Eastern Telegraph Co. Limited, minutes of the board, January 5, 1926, Minute Book No. 18, Cable and Wireless Archives.

44. "Eastern Telegraph Company," *Times* (London), May 25, 1927, 24. Denison-Pender was still making the same point in December 1929: Cable and Wireless Limited, *The Cable and Wireless Communications of the World: Some Lectures and Papers on the Subject, 1924–1939* (Cambridge: W. Heffer and Sons, 1939), 37.

45. "Competition between 'Beam' Wireless and Cable Service. Memorandum by General Post Office [F. W. Phillips]," July 6, 1927, Post 33/2177, M16773/1927, file 5, BT Archives.

46. Eastern Telegraph Co. Limited, minutes of the board, January 8, 1924, January 18, 1927, January 31, 1928, January 29, 1929, Minute Book No. 18, Cable and Wireless Archives.

47. Eastern Telegraph Co. Limited, Eastern Extension, Australasian and China Telegraph Co. Limited, minutes of the board, January 20, 1927, February 16, 1928, January 20, 1929, Minute Book No. 12, Cable and Wireless Archives.

48. Discussed in Boyce, "Canada and the Pacific Cable Controversy," 73–74 and passim. The Post Office–owned Imperial Atlantic cables also began to lose a substantial amount of money at this time. See "Wireless and Cables: The Suggested Merger," *Times* (London), May 7, 1928, 17.

49. "Beam Wireless and Pacific Cable," *Times* (London), November 25, 1927, 15.

50. G. Whiskeard to Secretary, GPO, enclosing telegram from governor general to secretary of state for the dominions, June 11, 1927, Post 33/2177, M16773/1927, file 5, BT Archives. See also the warning from the New Zealand prime minister: J. G. Coates to Mackenzie King, August 4, 1927, Papers of William Lyon Mackenzie King, vol. 166 (C-2295), p. 120757, University of Cambridge Library.

51. Minutes of fifty-fourth meeting of the I.C.C., July 11, 1927, Post 33/2177, M16773/1927, file 6, BT Archives.

52. CID Imperial Communications Committee, "Sub-Committee on Competition between 'Beam' Wireless and Cable Services," December 8, 1927, vol. 1247, file 30,039, pt. 2, RG 25, LAC.

53. Boyce, "Imperial Dreams and National Realities," 57–58 and passim.

54. Boyce, "Canada and the Pacific Cable Controversy," 76–84.

55. Sir Campbell Stuart to Mackenzie King, October 21, 1927, D1, vol. 750, file 191, RG 25, LAC.

56. November 23, 1927, CAB 23/55, Cabinet conclusion 57(27)15, NA; Leo Amery to Secretary of State for External Affairs, Canada, December 8, 1927, vol. 659, file 118164-1, RG 3, LAC.

57. Eastern Telegraph Company Limited, minutes of the board, October 25 and November 9, 1927, Minute Book No. 18, Cable and Wireless Archives; statement by prime minister, November 23, 1927, CAB 23/55, Cabinet conclusion 57(27)15, NA; Hugh Barty-King, *Girdle Round the Earth: The Story of Cable and Wireless and Its Predecessors to Mark the Group's Jubilee, 1929–1979* (London: Heinemann, 1979), 203–214.

58. November 23, 1927, CAB 23/55, Cabinet conclusion 57(27)15, NA.

59. January 16, 1928, CAB 32/65, I.W.C.(28), first meeting, NA.

60. December 7, 1927, CAB 23/55, Cabinet conclusion 60(27)2, NA.

61. Barty-King, *Girdle Round the Earth*, 205. Barty-King dates the offer to December 28, but according to company records, Denison-Pender was "brought in contact with Lord Inverforth"—and secured agreement in principle to a merger—several weeks earlier than this. Eastern Telegraph Company Limited, minutes of the board, December 6, 1927, Minute Book No. 18, Cable and Wireless Archives.

62. "Marconi's Wireless Telegraph Company," *Times* (London), March 16, 1927, 23; "The Marconi Report," *Economist* (March 12, 1927), 516–517; Baker, *A History of the Marconi Company*, 226–228.

63. See for instance the criticism of the Post Office in "The Telegraph Service," *Times* (London), March 24, 1928, 11.

64. "Marconi's Wireless Telegraph Company," *Times* (London), July 15, 1930, 21.

65. "City Notes," *Times* (London), March 20, 1928, 23; Barty-King, *Girdle Round the Earth*, 205.

66. January 31, 1928, CAB 32/65, I.W.C.(28), sixth meeting, NA; "Cable and Wireless Services, Plea for Competition," *Times* (London), March 22, 1928, 11; "Wireless v. Cables," *Economist*, January 7, 1928, 8; "Cables and Wireless," *Times* (London), November 30, 1927, 10.

67. Sobel, *ITT*, 63. It appears that the only significant takeover proposal made at this time was Carlton's offer to purchase the Pacific Cable Company. See S. J. Goddard, Western Union Telegraph Company, to Sir Campbell Stuart, January 30, 1928; Goddard to F. J. Edye, Secretary, Pacific Cable Board, February 10, 1928, vol. 657, file 117942, pt. 3, RG 3, LAC.

68. Memorandum from Lord Inverforth and F. G. Kellaway, February 9, 1928, CAB 32/67, I.W.C. (28)47, NA.

69. "Financial Notes," *Times* (London), March 22, 1928, 22.

70. Western Union had embarked upon a new Pacific cable linking San Francisco with Hawaii, Midway, Guam, and the Philippines, and was soon in financial trouble. Sobel, *ITT*, 62–63.

71. Sobel, *ITT*, 58–59.

72. R. W. D. Boyce, *British Capitalism at the Crossroads, 1919–1932: A Study in Politics, Economics, and International Relations* (Cambridge: Cambridge University Press, 1987), 181–184.

73. Thomas Jones, in *Whitehall Diary*, vol. 2, *1926–30*, ed. Keith Middlemas (Oxford: Oxford University Press, 1969), 177.

74. February 15, 1928, CAB 23/57, Cabinet conclusion 7(28)3, NA.

75. Amery to Mackenzie King, March 31, 1928, vol. 657, file 117942, pt. 3, RG 3, LAC.

76. G. E. P. Murray to Postmaster General, n.d. [ca. January 10, 1930], Post 33/5301 (Part 1), M20115/1938, file bottom B, BT Archives.

77. Statement by Sir John Gilmour, January 16, 1928, CAB 32/65, I.W.C. (28) first meeting, NA.

78. February 17, 1928, CAB 23/57, Cabinet conclusion 9(28)2, NA.

79. March 21, 1928, CAB 23/57, Cabinet conclusion 15(28)1, NA; April 18, 1928, Cabinet conclusion 21(28)5, NA; April 30, 1928, Cabinet conclusion 25(28)1, NA.

80. Stuart to Sir Norman Leslie, secretary to conference, February 15, 1928, vol. 657, file 117942 f. 3, RG 3, LAC; Stuart to Mackenzie King, March 23, 1928, Papers of William Lyon Mackenzie King, vol. 189, University of Cambridge Library; Stuart to O. D. Skelton, April 15, 1928, D1, vol. 750, file 191, RG 25, LAC; Stuart to L. J. Gaboury, postmaster general, May 2, 1928, file 117942, pt. 3, RG 657, LAC.

81. January 26, 1928, CAB 32/65, I.W.C. (28), fourth meeting, NA; March 21, 1928, CAB 23/57, Cabinet conclusion 15(28)1, NA.

82. Stuart to Sir Norman Leslie, February 15, 1929, vol. 657, file 117942 f. 3, RG 3, LAC; Sir John Gilmour to Leo Amery, March 12, 1928, ibid.; April 18, 1928, CAB 23/57, Cabinet conclusion 21(28)5, NA; April 30, 1928, CAB 23/57, Cabinet conclusion 25(28)1, NA.

83. Report, July 2, 1928, CAB 32/67, I.W.W.C.(28)75, NA; also published in "Imperial Cable and Wireless Conference, 1928," Report, *Parliamentary Papers, 1928*, Cmd. 3163, vol. x, pp. 601–623. The agreement between the companies and interested governments to form the merger company is in I&IC Contracts, Agreements &c. no. 1, May 29, 1929, Cable and Wireless Archives.

84. "Imperial Communications," *Times* (London), July 28, 1928, 13; "Imperial Telegraphs Bill," *Times* (London), November 22, 1928, 15.

85. Canada, which could count on interested assistance from its neighbor, the United States, in the event of foreign aggression, did not regard the cables or indeed the rest of the empire to be essential to its defense and security. See for instance L. J. Gaboury to O. D. Skelton, August 15, 1932, vol. 1502, file 938, pt. 5, RG 25, LAC; "Meeting at Department of External Affairs," minute by Skelton, October 18, 1932, ibid.; Secretary of State for External Affairs, Canada, to Secretary of State for the Dominions, Britain, November 9, 1937, DD1, vol. 750, file 193, RG 25, LAC.

86. "Wireless v. Cables," *Economist*, January 7, 1928, 8; Barty-King, *Girdle Round the Earth*, 207.

87. ICAC minutes of fifteenth meeting, March 18, 1930, I&IC Papers Submitted to Court Meetings, 1929–1930, Cable and Wireless Archives.

88. ICAC Minutes of sixteenth meeting, April 15, 1930, I&IC Papers Submitted to Court Meetings, 1929–1930, Cable and Wireless Archives; C. L. Baillieu to Sir Norman Leslie, secretary, ICAC, June 27, 1932, vol. 993, file 21-5-1, ICAC document 390-A, RG 3, LAC.

89. ICAC minutes of twenty-second meeting, October 13, 1930, I&IC Papers Submitted to Court Meetings, 1929–1930, Cable and Wireless Archives.

90. ICAC minutes of thirty-fifth meeting, July 8, 1931, I&IC Papers Submitted to Court Meetings, 1931, Cable and Wireless Archives. See also Lord Inverforth telegram to Sir John Higgins, June 12, 1933, I&IC Papers Submitted to Court Meetings, 1933.

91. ICAC minutes of forty-seventh meeting, March 31, 1932, vol. 993, file 21-5-1, RG 3, LAC; India Radio Telegraph Company to ICAC, March 30, 1933, vol. 993, ICAC document 366-A, RG 3, LAC.

92. R. H. Brand to Leslie, October 5, 1931, vol. 992, file 21-5-1, vol. 2, ICAC document 287-A, RG 3, LAC.

93. "Canada: Negotiations with Canadian Marconi for Combined Working of Cables and Wireless Systems," n.d., I&IC Papers Submitted to Court Meetings, 1931, Cable and Wireless Archives; James Fergus, Secretary, Canadian Marconi Company, Montreal, to secretary to the Post Office, London, May 18, 1931, ibid.

94. Edward Wilshaw, General Manager and Secretary, I&IC Ltd., to Chairman, RCA, February 29, 1932, vol. 993, file 21-5-1, ICAC document 350-A, RG 3, LAC.

95. Wilshaw to Leslie, March 21, 1932, vol. 993, file 21-5-1, ICAC document 365-A,

RG 3, LAC; Minutes of forty-seventh meeting, March 31, 1932, ibid.; draft Tripartite Agreement, June 22, 1932, ibid., ICAC document 388-A; A. C. Baillieu to Leslie, June 27, 1932, ibid., ICAC document 390-A. See also W. M. Jardine, Cairo, "strictly confidential," to Secretary of State, no. 450, April 15, 1932, 883.73/1, RG 59, NARA; Jardine to Secretary of State, no. 649, January 27, 1933, ibid., 883.73/9.

96. Wilshaw to Leslie, July 5, 1932, vol. 993, file 21-5-1, ICAC document 394-A, RG 3, LAC; "Wireless Station at Shanghai," I&IC Ltd memorandum, July 29, 1932, ibid., ICAC doc. 407-A.

97. Ibid., Wilshaw to Leslie, March 9, 1932.

98. ICAC minutes of twenty-sixth meeting, December 16, 1930, I&IC Papers Submitted to Court Meetings, 1929–1930, Cable and Wireless Archives.

99. See the warning to this effect by J. Gosling, "Wireless and Cables," *Times* (London), July 2, 1928, 12.

100. ICAC minutes of sixteenth meeting, April 15, 1930; ICAC minutes of eighteenth meeting, June 24, 1930, I&IC Papers Submitted to Court Meetings, 1929–1930, Cable and Wireless Archives.

101. Agreement, February 19, 1932, vol. 993, file 21-5-1, ICAC document 355-A, RG 3, LAC.

102. Memorandum by Sir Basil Blackett, "Telegraphic Communications of Canada," January 15, 1931, vol. 1594, file 178, RG 25, LAC.

103. Note by the secretary, June 19, 1931, vol. 1502, file 938, pt. 5, ICAC document 257-A, enclosure no. 2, RG 3, LAC.

104. Memorandum, "The Imperial and International Communications Company Limited and the Canadian Marconi Company," July 9, 1931, ICAC document 259-A, RG 3, LAC; Minutes of the thirty-sixth meeting, July 29, 1931, vol. 992, file 21-5-1, vol. 2, RG 3, LAC.

105. Sir J. Flavelle to Hon. A. Durandeau, June 6, 1931, vol. 1502, file 938, pt. 5, RG 25, LAC; Ferguson to R. B. Bennett, July 27, 1931, idid.; Assistant Deputy Postmaster General to Skelton, May 28, 1936, vol. 658, file 118164-B-1, RG 25, LAC.

106. Lyons to Amery, February 27, 1929, Stuart Papers, vol. 6; statement by Lyons, *Telegraph and Telephone Journal*, February 1931, 100.

107. Estimated Message Receipts, Month of June 1929, I&IC Papers Submitted to Court Meetings, 1929–1930, Cable and Wireless Archives. The six-month figures were £3,282,044 for 1928 and £3,139,664 for 1929.

108. Estimated Message Receipts, December 1930, I&IC Papers Submitted to Court Meetings, 1931, Cable and Wireless Archives.

109. Blackett to F. B. Lees-Smith, postmaster general, July 27, 1929, Post 33/5301 (Part 1), M20115/1938, file bottom D, BT Archives; Cabinet Wireless Telegraph Committee

Report, 19.2, App. B, stenographic notes of evidence, second meeting, November 25, 1929, file bottom E, W.T.C. (29)24, BT Archives; "Dissent of GPO Representative," December 18, 1929, BT Archives; Post Office memorandum, October 20, 1932, Post 33/5301 (Part 2), M20115/1938, file 16, BT Archives; Post Office memorandum, September 3, 1935, file 18, BT Archives. The company's campaign was strongly supported by the conservative London press. See leader, *Financial Times*, February 12, 1930; interview with Sir Edward Hilton Young, ibid., February 14, 1930; leader, ibid., March 1, 1930; "Wireless and the Post Office," *Times* (London), February 27, 1930, 15; "The Cables and Wireless Problem," *Financial News*, October 27, 1932, 5.

110. Secretary of State for the Dominions, Britain, to Secretary of State for External Affairs, Canada, March 24, 1931, vol. 992, file 21-5-1, vol. 2, RG 3, LAC; Report (n.d.), Stuart Papers, vol. 12, ICAC No. 1 (1931).

111. "Report to the Governments of the Empire, relating to the Committee of Inquiry composed of Mr. Wilfrid Greene, The Rt. Hon. Lord Ashfield, and Mr. Lawrence D. Holt," November 21, 1931, CAB 21/485, NA. The three-man report is contained in Appendix 1. A list of the cables that I&IC Ltd wished or might wish to supersede is contained in Appendix 4. A copy of the report is in the Stuart Papers, vol. 9, and another in vol. 1502, file 938, pt. 5, RG 25, LAC. On the directors and their remuneration, see also High Commissioner, Canada, to Secretary of State for External Affairs, Canada, March 25, 1929, vol. 659, file 118164-2, RG 3, LAC.

112. "Second Report to the Governments of the Empire relating to the Report of the Committee of Inquiry," November 27, 1933, Stuart Papers, vol. 9, ICAC doc. 9-B; "Third Report to the Governments of the Empire relating to the Report of the Committee of Inquiry," May 1, 1934, ibid., ICAC doc. 10-B.

113. Barty-King, *Girdle Round the Earth*, 240. At the same time the holding company also changed its name from Cables and Wireless Ltd. to Cable and Wireless (Holding) Ltd.

114. CAB 21/842, Report of the Imperial Communications Advisory Committee, January 21, 1937, NA.

115. Report to the Governments of the Empire, November 30, 1937, Stuart Papers, vol. 9, ICAC doc. 14-B; also published in "Cable and Wireless Limited," United Kingdom, *Parliamentary Papers, 1937–1938*, Cmd. 5716, vol. xxi, pp. 823–829, and Sir Campbell Stuart, *Opportunity Strikes Once* (London: Collins, 1952), appendix 2.

116. Report of the Imperial Communications Advisory Committee, January 21, 1937, CAB 21/842, NA.

117. Strategic Cables Sub-Committee, June 30, 1932, Post 33/3779, M13737/1932, file 2, ICC(S.C.)1. CID. ICC, BT Archives; "Strategic Cables. Note by Secretary," April 17, 1937, file 4, Secret I.C.C. No. 1738, CID. ICC, BT Archives.

118. Strategic Cables Sub-Committee, "Summary of conclusions by the Service Representatives at a meeting held on 19 October 1932," Post 33/3779/1932, file 2, ICC(S.C.)7. CID. ICC, BT Archives; "Strategic Cables," minutes of meeting on April 22, 1937, file 4, BT Archives.

119. The 1928 rates are set out in "Imperial Wireless and Cable Conference, 1928. Report," United Kingdom, *Parliamentary Papers, 1928*, Cmd. 3163, vol. x, p. 7. The new rates and estimated savings are given in "Cable and Wireless Limited," United Kingdom, *Parliamentary Papers, 1937–1938*, Cmd. 5716, vol. xxi, pp. 5, 6.

120. "Evidence given by Sir Basil Blackett on the 25th November, 1929, before the special committee appointed by the Cabinet to consider the question of overseas telephony," I&IC Papers Submitted to Court Meetings, 1929–1930, Cable and Wireless Archives.

121. Baker, *A History of the Marconi Company*, 270–271; Barty-King, *Girdle Round the Earth*, 236, 249.

122. Wilshaw to I&IC Ltd., September 19, 1930, App. II, I&IC Ltd., "Report to the Governments of the Empire," September 1930, Stuart Papers, vol. 9; Report to the Governments of the Empire, May 19, 1936, Stuart Papers, ICAC doc. 866-A.

123. Secretary of State for External Affairs, Canada, to Secretary of State for the Dominions, Britain, November 9, 1937, D1, vol. 750, file 193, RG 25, LAC; "Australia-USA Radiotelephony Service," January 18, 1939, Stuart Papers, vol. 19, ICAC document 1092-A; Barty-King, *Girdle Round the Earth*, 251–252 and passim.

124. Boyce, *British Capitalism at the Crossroads*, 366–369 and passim.

125. Sobel, *ITT*, 65; Headrick, *The Invisible Weapon*, 209.

126. Ibid.; Ludwell Denny, *America Conquers Britain: A Record of War* (New York: Alfred A. Knopf, 1930), 396–399; Case and Case, *Owen D. Young and American Enterprise*, 483. The same argument was successfully used to justify the exemption that allowed the sale of the (U.S.) Postal Telegraph Cable Company to the Commercial Cable Company ten years later. See W. R. Vallance, State Department, to Howard L. Kern, General Counsel, Postal Telegraph Cable Co., the Mackay Companies and the Commercial Cable Companies, March 8, 1938, box 5226, file 811.73 P84/10, RG 59, NARA. The Federal Communications Commission approved the sale on February 13, 1939.

127. G. Stanley Shoup, "Wireless Communications in the British Empire," Trade Information Bulletin no. 551 (U.S. Department of Commerce, Bureau of Foreign and Domestic Commerce, 1928); Minute by Irwin Stewart, October 23, 1933, box 5218, file 811.70/8, RG 59, NARA; Captain S. C. Hooper to Chairman, President's Communications Policy Committee, December 7, 1933, ibid., file 811.70/11; Claude Swanson to Cordell Hull, September 15, 1934, enclosing Report of Joint Army-Navy Board of January 19, 1934, ibid., file 811.70/29; "Projected Merger of Communications Companies in the United States of America and its Effects on Empire Communications," February 26, 1934, Stuart Papers, vol. 8, ICAC document 591-A; Denny, *America Conquers Britain*, 367–402.

128. Stuart, *Opportunity Strikes Once*, 122–126 and appendix 2; Barty-King, *Girdle Round the Earth*, 286–287 and passim.

7

Small Nation, International Submarine Telegraphy, and International Politics: The Great Northern Telegraph Company, 1869–1940

Kurt Jacobsen

On January 31, 1930, Alfred Schønebeck, the Moscow representative of the Great Northern Telegraph Company (GNTC), appeared before the supreme Soviet Planning Commission, Gosplan. The GNTC was a Danish-owned, international cable company operating in Europe and the Far East, and it remains unclear why a foreigner like Schønebeck was given access at all to this assembly of Stalin's top economists, but it undoubtedly happened on his initiative. His objective was to present a discourse on the Danish firm and its activities and to assure the Soviet authorities that cooperating with the GNTC meant both security and other benefits for the Soviet Union. In this connection he stressed, first and foremost, the company's political neutrality:

> An international telegraph company such as ours, which works in 10 different states and has links to an even greater number of governments and international organizations, has naturally first and foremost to observe the strictest neutrality in all political matters and absolute loyalty toward all the governments with which it collaborates.[1]

Second, he stressed the operational advantages of the company's telegraph system as constituting an "administrative and technical unity":

> In this case the operating of the whole system, across national and geographic boundaries as well as local habits, is regulated from one single center, that, without protracted negotiations with other parties, has the authority and the capability to determine the most appropriate rules for the handling of the traffic.

Third, he stressed the political-administrative and diplomatic advantages of letting the GNTC be responsible for the Soviet Union's international telegraph links:

> In practice, achieving the best operation of international lines demands, when national boundaries are involved, an extensive and complicated official negotiating apparatus to be set in motion, and the more state

> administrations involved in the discussion, the more difficult it is to organize all the practical details that together make for a good telegraph service.

Schønebeck's arguments, of course, were dictated by the situation, but they entailed much more than that. In fact, they expressed the very foundation of the GNTC's existence as an international telegraph company for more than sixty years—reflecting the fundamental characteristics of international telegraphy and its business environment. In this context, Schønebeck's first argument hinted at the strong political interests that, from the very beginning, played a prominent role in the business of international telegraphy. His second argument pointed to the nature of the industry itself, based as it was on a large technical system. And his third argument dealt with the special combination of the two characteristics above that constituted the unique business environment in which the GNTC operated. Those characteristics had made it possible for the GNTC to establish a leading position in an industry dominated by a handful of big and powerful western European, and especially British, companies.

A systematic analysis of the business environment and its constituent characteristics is therefore necessary in order to explain the development of the GNTC and of the industry as a whole. At the same time, tracing the development of the GNTC will help to clarify the business environment that determined the evolution of international telegraphy.

International Regulation and Private Cable Companies

The telegraph emerged as an archetype of the kind of systems known as "large technical systems" (LTS), the capital-intensive organizations that were established from the second half of the nineteenth century to provide solutions to the crucial needs of the modern industrial society: transportation (railroads), telecommunications (telegraphy, telephone), energy supply (gas, electricity), and different kinds of urban utility networks (water supply, sewers). Because of their obvious technical character, LTSs have traditionally been viewed as "the domain of engineers," having limited social relevance, but since the publication in 1983 of Thomas Hughes's *Network of Power* they have been viewed as complex sociotechnical organizations that emerged and developed in close coevolution with "society at large."[2] In the case of international submarine telegraphy the LTSs were organized as business organizations with the purpose of earning their owners a profit.

Because of their essential importance to modern society and their character of being territorially organized, or "natural," monopolies, the LTSs were always placed under close public control and regulation. Often they were owned and operated by local authorities or by the state—the latter being true in the case of the telegraph. With the United States as the main exception, government administrations all over the world established and operated the national telegraph networks, while already existing private networks were nationalized early on.

With its fast expansion in Europe, telegraphy soon became international. Technically there were no major problems in linking the different national telegraph networks, though differences in equipment and standards created some minor obstacles. It involved, however, the creation of a functional system across political borders and the establishment of cooperation between government administrations. Because of this, international telegraphy became a matter of international politics.

The first treaty on record to link the telegraph networks of two states was signed on October 3, 1849, by Prussia and Austria. It soon was followed by similar agreements between Prussia and Saxony and between Austria and Bavaria, and on July 25, 1850, the four states signed a convention, creating the German-Austrian Telegraph Union. This central European example was followed in June 1855 by the formation of the West European Telegraph Union, bringing together Belgium, France, Sardinia, Spain, and Switzerland.[3]

The provisions in the two conventions were almost identical and, in October 1852, the two groups became connected when Belgium, France, and Prussia signed a mutual telegraph convention. These three countries were the core in a development that, on May 17, 1865, led to the formation of the International Telegraph Union (ITU). The convention was signed in Paris by 20 European states. Of the important countries of Europe, only Great Britain was absent. The world's most powerful nation had not been invited to join the convention because her telegraph services were operated by private companies; but after their nationalization in 1870, Great Britain too became a member.[4]

With the Paris convention, common rules were brought to bear on international telegraphy, together with a set of regulations on technical standards, codes, and services. A table of tariffs appended to the document introduced uniform telegram rates for all the signatory countries, with the French franc as the common monetary unit. The convention and regulations came into effect on January 1, 1866, and in the following years ITU was joined by most countries in the world. The United States remained an exception because American telegraph companies were never nationalized. The American government, however, accepted the status of observer in the ITU and agreed to follow its rules and regulations. Thus in the last third of the nineteenth century, the telegraph system developed into a global network under close regulation and government control on both national and international levels. In this process, the ITU had been established by governments to connect their national networks, and its political character was reflected by its being "the first genuine, intergovernmental organization to see the light of the day."[5]

The convention, however, also had to take into account that international telegraphy no longer was solely a government affair. With the successful laying of the first submarine cable across the English Channel in 1851, private companies entered international telegraphy and, whereas governments continued to establish and operate their national networks, the establishment and operation of overseas connections was usually left to private enterprises. This happened partly for financial reasons, for it was very risky and costly to manufacture and lay submarine cables, and partly for political reasons. Collaborating on financing, constructing, and operating overseas cables was very different than linking landlines—even though this occurred over short distances as, for instance, with the Danish-Swedish submarine cable that was laid by the two governments in November 1855.[6]

In consequence, the development of overseas telegraphy was linked to the rise of a powerful cable and telegraph industry, especially in Great Britain. Being private, the cable companies were not directly subordinated to the ITU, but the Paris convention instructed the contracting governments to impose the convention's regulations on all private companies operating under their authority. In 1868 private telegraph companies were permitted to accede to the convention with a notification from their home country, and in 1871 private companies were admitted to the ITU conferences with the right to participate in discussions, but without the right to vote.[7] Denmark was one of the founding members of the ITU, and the Great Northern Telegraph Company was represented at all its conferences after 1871.

International telegraphy thus developed a very complex structure, founded on the continuing collaboration between governments and private enterprises. Each of the international telegraph companies had the character of an LTS based on the private ownership of submarine cables that were linked to other cable networks and to the national networks. As private enterprises they operated in a business environment that was under heavy political influence and characterized by close government surveillance and regulation, and by the direct participation of state administrations in the business field.

It was exactly these characteristics that Schønebeck pointed out to Stalin's top economists: the GNTC possessed the essential features of an LTS that "across national and geographic boundaries" operated "from one single centre." A telegram sent from Shanghai "via Northern" to London would have to pass through not only the GNTC's cables in Europe and the Far East, but also the territories of China, Japan, Soviet Russia, Finland/Sweden or Latvia/Denmark, and finally Great Britain—with the GNTC as the operator responsible. In this respect, the GNTC was unique as an LTS since its telegraph system included several national telegraph lines, but this special feature reflected the general complexity and the political environment of international telegraphy.

Private Cables and Government Intervention

It was one thing, however, that international telegraphy became a political issue tied up with international cooperation and regulation; it was quite another that the new means of communication, as pointed out by Jorma Ahvenainen, fundamentally changed world politics, into which, "more than any other means of communication, the telegraph introduced the time factor."[8]

The first incident in which the international telegraph revealed its military and diplomatic implications took place during the Crimean War in 1854, when a cable was laid from Varna on the Romanian Black Sea coast to the Crimean peninsula. For the very first time, French and British governments could communicate directly with their commanders on a distant battlefield. At the same time, to the displeasure of the governments, newspaper correspondents were sending news to Paris and London about the chaotic nature of British-French warfare. Furthermore, daily reports of the war, and of the military plans, could be telegraphed to Moscow the very same day that they appeared in French and British newspapers.[9]

Another incident occurred during the war between Prussia/Austria and Denmark in 1864. When Prussian troops occupied the southern duchies—Schleswig, Holsten, and Lauenborg—they also captured the Danish landing spot of the submarine cable to Great Britain, with the effect that Denmark lost the direct telegraphic connection to London that had been established in 1859. The Danish government was therefore forced to use secret couriers traveling through the Prussian-occupied territories in order to communicate with its ambassador in London, a practice that was both slow and risky.

While the Crimea incident had demonstrated both the political and military significance of international telegraph links as well as the importance of controlling the flow of information that passed through them, the Danish incident demonstrated the possible adverse consequences to a country of losing an already established, international telegraph connection. In both cases, the implications for foreign affairs and national security policy were obvious, and even though submarine cables were laid and operated by private enterprises in a "seemingly apolitical atmosphere," as Daniel Headrick puts it, they nonetheless came into the focus of government politics for other reasons than regulation.[10]

This brings us back to Schønebeck and his appearance before Gosplan in January 1930, brought about by a substantiated fear of politically motivated action against the company and its activities. The Soviet Union had, from the late 1920s, embarked upon the road toward industrialization under the first five-year plan—a process that was accompanied by ideological and nationalistic rhetoric against capitalism and the surrounding world. As a consequence, foreign companies were facing increasing difficulties, as they were put under pressure by the Soviet government to leave the country. Furthermore, foreign specialists had appeared in the dock during the first show trials of the late 1920s, followed by hard-hitting propaganda against spies and persons from abroad.

Schønebeck and the GNTC management in Copenhagen were fully aware that sooner or later the company would most probably become an object of scrutiny by the Soviet authorities and would share the fate of other foreign companies. Instead of waiting for a Soviet action against the company, Schønebeck had chosen to take the initiative, knowing that his performance might determine the company's future. The situation was in every respect extreme, and the threat correspondingly dramatic.

Apart from the special circumstances caused by the nature of the Soviet regime, however, the situation was not unusual. Rather, it reflected the general business conditions under which the GNTC had operated since its foundation in 1869. Throughout the company's history, governments had shown a constant and growing concern regarding the control of submarine cables, irrespective of the fact that these were private property. It was a concern—deeply rooted in the Crimean and the Danish-Prussian cases as well as others—reflecting the view that it was increasingly essential that a country control its international telecommunications. In this context, the Soviet threat to the GNTC was just another potential government intervention in its business, although undoubtedly it was among the more serious ones.

The Political Imperative in International Telegraphy

Not only was the GNTC under constant threat of government interventions, but, as a private company, it managed its cables and network as an

LTS under close public supervision and regulation. In order to explain the GNTC's position it is necessary to analyze the business conditions of international telegraphy. It will therefore be useful to introduce a conceptual framework from international business theory that accounts for interdependencies between the firm and its political environment.

To describe the different kinds of government intervention that multinational firms confront in doing business internationally, Peter Smith Ring, Stefanie Ann Lenway, and Michele Govekar use the term *political imperative*, which they understand primarily as political risk, even though they are aware of the other basic understanding of the term as political opportunity. They define political risk as "home or host government intervention in international business," with a focus on the "negative effects on the firm, a result that is consistent with research on how managers view political risk."[11]

As argued by Ring, Lenway, and Govekar, the impact of the political imperative on the firm varies with the industry structure: "Within specific industries, some strategies may be more effective than others because they help to insulate the most vulnerable aspects of the firm from government intervention." In their concept of industry structure they distinguish between "global" and "multi-domestical" industries, based on the assumption that industry structure is either a consequence of economic imperatives or based on transaction cost.[12]

This concept, however, is not directly applicable to international telegraphy because of the essential features of LTSs that dictated the structure of the industry as well as that of the individual firms. Nevertheless, international submarine telegraphy was indeed international business and as such was considered by governments to be an industry of strategic importance.

An international telegraph company like the GNTC was, of course, an LTS whose purpose was to transmit telegrams from senders to recipients. From a technical point view this was a simple operation to manage; however, it had to be done under close public control and supervision in accordance with the ITU's convention and regulations and, at the same time, in direct collaboration with a number of government administrations. On top of these "ordinary" conditions for doing LTS business came the more unpredictable possibility of government intervention caused by the national and international political impact of the new means of communication.

In international telegraphy, the political imperative was present everywhere at all times. It was this politically dominated business environment that came to determine both the establishment and the long-time development of the GNTC. Ring, Lenway, and Govekar rightly state that "the failure of firms to adapt to the political demands of a specific industry structure is a consequence of the divergence between the strategy that is responsive to the political imperative within a specific industry structure and the strategy that firms adopt as a result of their strategic predisposition."[13]

In the case of the GNTC, there was no such divergence. The company realized from the very beginning that the management of submarine cables was not merely the management of an LTS; in particular, it entailed management of the political imperative, meaning both politically determined opportunities and risks. The strategy of the GNTC was deliberate and consistent. It was designed—as stated by Schønebeck at his appearance before Gosplan in January 1930—to meet any political challenges on the basis of "the strictest neutrality." It involved a politically and diplomatically defined predisposition created to respond to political opportunities that led to both the formation and the rapid expansion of the company. But it was also an approach that gradually lost its potency and effectiveness with the coming of the twentieth century and the radical changes in world politics.

In the Service of the Nation: The Founding of the GNTC

The GNTC was founded on June 1, 1869, through a merger of three telegraph companies connecting Denmark, Norway, Sweden, and Russia through submarine cables. The merger was the final result of a deliberate business strategy that took advantage of Denmark's geographical location, its small state status and neutrality, and the nation's desperate need for telegraphic connections to Great Britain and Russia.[14]

The formation of the company was the result of Denmark's loss, in 1864, of the direct cable to Great Britain that had left the Danish government in a politically unacceptable situation, since all its telegraphic correspondence with London had become dependent on a telegraph line that crossed German territory. At the same time, the loss resulted in decreasing transit incomes to the Danish state. Moreover, when a Baltic cable between Sweden and Prussia opened in June 1865, the financial situation became even more serious, since almost all telegrams between Norway/Sweden on the one side and Great Britain/continental Europe on the other now bypassed Denmark, resulting in significant economic losses for the Danish state.

The Danish government had asked several British companies, including the Telegraph Construction and Maintenance Company, to lay a new cable to Great Britain, but they had all declined, arguing that the telegram traffic between Great Britain and now diminished Denmark was not great enough to ensure an adequate return on the investment. Not even a proposal from the Danish government to the British companies offering to grant them a concession to lay and operate a cable between Denmark and Russia (and thus giving them access to the British-Russian telegraph traffic) made them change their position. They were, after all, already operating cables across the Channel and for obvious reasons did not want to compete with themselves.

Instead, a non-British company, Siemens & Halske of Berlin, showed interest in laying a cable to Russia. The Danish government was willing to grant a concession to the German company, but for political reasons the Russian government was reluctant. Telegraphic connections between Russia and Great Britain already existed via the continental landlines and the cables across the Channel. Russia, however, wanted an alternative, and in March 1866 the Russian minister of the interior informed the Danish Foreign Ministry that his government wanted a "direct link to Great Britain, independent of Germany." The country did not have the industrial or the financial capacity to establish the link itself, so it had to turn to foreign companies.[15]

The political nature of the Russian motives made it impossible to deal with a German company. Faced with the reluctance of the British companies, the Danish government turned to the Danish financier and director of the Private Bank in Copenhagen, C. F. Tietgen. For several years Tietgen had acted as the Danish representative for British telegraph interests, and in 1867 he had assisted the British cable manufacturer R. S. Newall in laying a cable between Norway and Denmark.

The main problem for Tietgen was financial. Besides the laying of a cable between Denmark and Great Britain, he wanted to take over the Danish-Norwegian cable from R. S. Newall, a goal that made it necessary to raise £100,000—and on top of this came the financing of the cable to Russia. The task seemed almost impossible, considering the size of the Danish money market and the difficult economic situation after the loss of the southern duchies, but Tietgen took advantage of the interests of the Danish state and persuaded the government to invest £18,000 in the cable to Great Britain. Therefore, when the Danish-Norwegian-British Telegraph Company announced on January 10, 1868 a share issue worth £100,000, 18 percent had already been taken by the Ministry of Finance. This contributed to an unexpected rush for shares. With the permission of the Ministry of Finance, Tietgen was able to transfer the £18,000 to the Danish-Russian project, while the cable to Britain was laid in September 1868.[16]

Tietgen was well aware of the Russian desire for a telegraph link to Britain that bypassed Prussia and instead used Denmark as a neutral stepping stone. He therefore persuaded the Russian government to accept a 20 percent reduction in Russian telegram rates, which had been fixed by the

ITU, alongside an arrangement whereby all telegrams from Russia to Britain should be sent through the Danish cables and not via the continental landlines. The agreement would result in economic losses to the Russian state. But in a report to the czarist government, the Russian telegraph director emphasized instead that the country would gain an "independent" telegraph link to Great Britain, which would be of great importance in the event of a political crisis, or even war, between France and Germany.[17]

The importance to the Danish state of the establishment of the telegraph link was demonstrated by the intervention of the Danish king. Relations between Denmark and Russia were friendly after the marriage of the Danish princess, Dagmar, to the successor to the Russian throne, Grand Duke Alexander, in 1866. Tietgen's position was therefore strengthened when the Danish king, during a royal visit to St. Petersburg, recommended the project, and in September 1868 the Russian government granted the financier a monopolistic concession to lay the cable. A Danish-Russian Telegraph Company with a share capital of £75,000 had already been set up, and in May 1869 the cable between Russia and Great Britain via Denmark was ready to be opened—to the great satisfaction of the Danish as well the Russian government.[18]

On June 1, 1869, Tietgen founded the Great Northern Telegraph Company by an amalgamation of the Danish-Norwegian-British and the Danish-Russian companies together with a Norwegian-British Telegraph Company, which had been founded in Oslo to lay a cable between Norway and Scotland. As Tietgen at the same time obtained the necessary concessions to lay a cable between Finland and Sweden, the GNTC was in control of all submarine cables between Denmark, Norway, Sweden, Britain, and Russia except for the government cable between Denmark and Sweden. The system was extended in 1873, when a direct cable was laid between Denmark and France, thus providing Russia with a direct telegraph link to Paris that bypassed Great Britain as well as Germany.

Because of its geographical position and neutral political status, Denmark had become a hub in international telegraphy. Simultaneously, Tietgen had created a Danish-based, international telegraph company with a monopolistic and almost unshakeable position in northern Europe. His business strategy had been shaped by a situation where political imperative was equal to political opportunity, created through the British telegraph companies' reluctance to lay the requisite cables. By taking on this task, he had acted in the service of the Danish nation and also served the interests of the Russian government: the Danish and Russian governments both got the cables they desired.

While the motives of the Danish government had been political and financial (which also involved politics in so far as state finances were concerned), the Russian motives had been entirely political and strategic, based as they were on the country's great power interests. Thus it was the political imperative that helped Tietgen obtain the Danish and the Russian concession and the financial support from both governments. Furthermore, it was the political imperative that made it possible for Tietgen to break the British dominance of international telegraphy and to challenge Sir John Pender and his powerful group of associated companies that was already being formed around the Telegraph Construction and Maintenance Company. This group had done its best—but without success—to hinder the Danish financier in his venture, including attempting to sabotage the manufacture of the Danish cables by independent British cable producers. Tietgen was warned that British telegraph interests would not "permit these Danes to go further."[19] But they did, once more taking advantage of Denmark's geographical position and status as a small, neutral state.

From a Small, Neutral Country

Tietgen had a vision of making Denmark a hub for international telegraphy on a global scale. He planned a cable connection to North America via the Danish possessions of the Faroes, Iceland, and Greenland, as well as a link to Japan, China, and Hong Kong via Siberia. It was a vision based on the conscious exploitation of Denmark's neutrality and the expectation that, in the event of conflict between the great powers, international traffic would continue via lines operated by a neutral, Danish company.[20]

While the North Atlantic plan was abandoned because of difficulties in acquiring the necessary capital after the successful laying of the transatlantic cables in 1866, Tietgen succeeded in his ambitions for a link to the Far East. The key was a Russian concession to lay and operate cables from Vladivostok to Nagasaki, Shanghai, and Hong Kong, which he was awarded in October 1869 in keen competition with three other consortia. All the competitors represented British telegraph interests—including Sir John Pender and the Telegraph Construction and Maintenance Company.[21]

Again the Russian interests were purely political and strategic. For military and security purposes, the government wanted to extend an already existing landline from Moscow to Stretinsk, near Lake Baikal, further on to Vladivostok, where it had been decided to build a naval harbor. The government itself lacked the financial resources to extend the link from Stretinsk to Vladivostok but hoped it could be financed through transit revenues from trans-Siberian telegraph traffic between Europe and the Far East. For this purpose it needed a partner in the form of a private cable company that could lay the cables in the Far Eastern waters.

Tietgen was again perfectly aware of the Russian motives. So, just as he had previously exploited the Russian interest in a direct telegraph link to Great Britain that bypassed Germany, he now played on the rivalry and conflict of interests between Russia and Great Britain caused by the expansion of the two empires into Central Asia and the Far East. In his application to the Russian government, which was overtly political in its formulation, he warned that the immediate consequence of granting the concession to a British consortium would be "the commercial and political supremacy of Great Britain in China and Japan."[22]

Tietgen's political analysis convinced the Russian Council of Ministers, where his views were supported by Grand Duke Alexander after further interference by the Danish king. He was granted a monopolistic concession to lay cables in the Far East and, at the same time, the Russian telegraph administration committed itself to building a telegraph line connecting Vladivostok with Stretinsk and the European telegraph network, thus creating a trans-Siberian telegraph link connecting China and Japan with western Europe.

When Tietgen gained the Russian concession, he had no idea at all whether he would be permitted to land submarine cables in Japan and China. He assumed, however, that it would be relatively easy for a company from a small country without imperial interests to obtain the necessary permits.

His assumption proved correct to the extent that the Japanese government, on September 20, 1870, granted him a concession to land telegraph cables in Japan and to erect and operate a telegraph station in Nagasaki. However, he did not ask the Chinese authorities for permission and landing rights since he was convinced that the request would be refused. The Chinese authorities feared the telegraph as a tool for foreigners to "penetrate the interior" of China.[23]

Tietgen's plan was fully supported by the Danish government, which strongly believed that the GNTC was acting in the nation's best interests. The pleasing prospect of increasing transit revenues from the Far Eastern–European telegraph traffic would certainly be to the benefit of the Danish state. Denmark did not, however, have its own diplomatic representation in the Far East; it was represented by the Russian foreign service, which also supported the Danish project. A special Danish envoy was therefore sent to Japan and China while the Ministry of Foreign Affairs sought the support of the other great powers. Furthermore, a frigate was placed at the disposal of Tietgen and on October 20, 1870, the first telegraph cable in the Far East was landed by a Danish naval ship in Deep Water Bay, Hong Kong. Less than two months later, submarine cables were illegally and surreptitiously landed at Shanghai and, on April 20, 1870, the first submarine telegraph link in the Far East was opened between Hong Kong and Shanghai. On January 1, 1872, the two cities were linked to Europe via Nagasaki, Vladivostok, and the trans-Siberian landline, and the GNTC had established itself as an intercontinental telegraph company.[24]

The opening of the Danish telegraph network caused only a modest protest from the Chinese authorities against the illicit landing of the cables. The reason was first and foremost Denmark's nonthreatening political status. A similar action by a British company would have undoubtedly caused a much more aggressive and determined reaction. In this, the GNTC took advantage of a split among the high-ranking Chinese officials. The powerful viceroy Li Hongzhang did not share the official hostility against the telegraph; on the contrary, he wanted to modernize the country. In fact, he seemed quite pleased that it was a Danish company that introduced the telegraph into China.[25]

Finally, Tietgen benefited from the interest of the European colonial powers and the United States in opening China up to Western trade and influence. Following the Opium Wars, the United States (in the Burlingame Treaty of June 28, 1868) and Great Britain (in the Clarendon Declaration of December 28, 1868) expressly declared that they would not impose on China such modern means of communication as railways and the telegraph. This put a damper on British telegraph companies' opportunities and operations. The GNTC, however, had no such promises to break. On the contrary, its actions at Shanghai met with common wishes of the great powers—a role that the company could only play by virtue of Denmark's neutral status. When the Chinese authorities made their modest protest against the illicit landing of the cables, the GNTC was backed by the diplomatic representatives of all the great powers until the Chinese authorities finally accepted the landing of the Danish cables.[26]

Again, the political imperative had been decisive and once again the Pender group, despite its superior economic and organizational strength, had lost a confrontation with the far smaller competitor. This time the battlefield had been far away from Denmark, but the GNTC had managed to establish itself in a region with substantial Great Power interests. After this success, however, the GNTC had to face the economic imperative resulting from the LTS characteristics of international submarine telegraphy.

System-Building and Strategic Alliance in the Far East

The successful expansion of the GNTC into the Far East took place when the general climate of global politics created favorable opportunities for a company from a small, neutral state. It was the age of empire, and the GNTC came to play the role of "tool of Empire," to use Daniel Headrick's term:[27] not a tool of a Danish or any other particular empire, but rather a tool of the "global empire," which as a whole benefited from the opening up of China. Still, this does not mean that the GNTC did not confront imperialistic interests.

After the defeat in the struggle for the Russian concession, Pender tried to sabotage Tietgen's attempt to raise capital in London, just as he had sent a special agent to the Far East with the task of hindering Tietgen's attempt to gain concessions and landing licenses in China, Japan, and Hong Kong. But Pender did not leave it at that. In autumn 1871, a company under the control of the Pender group established a submarine telegraph link from India to Hong Kong—but with no connection to Shanghai—laying the foundations of the mighty Eastern and Associated Telegraph Companies, formed in 1873. Everything had been lined up for a total confrontation, but Pender apparently had gained respect for his Danish competitor. In February 1873 he concluded a joint purse agreement with Tietgen, enabling both companies to use GNTC's cable between Hong Kong and Shanghai and, at the same time, regulating the traffic and setting tariffs in the Far East and between China and Europe.[28]

The Danish-British partnership, however, came to an end in 1881, when Li Hongzhang granted the GNTC a monopoly on China's international telegram traffic as well as the right to assist in China's development of the country's internal telegraph network. A monopoly was unacceptable per se to the great powers, and while they earlier had backed the GNTC, the Danish company this time received only the support of Russia, making matters even worse. Russo-British antagonism had worsened significantly following the two countries' conflict over the control of Afghanistan so, in British eyes, the GNTC represented Russian interests by virtue of its telegraph link via Siberia, which would now be the sole conduit for traffic between China and Europe.

After more than five years' war of attrition, the Danish company concluded a peace in December 1886 and entered a partnership agreement with the Eastern and Associated Telegraph Companies. The GNTC had suffered a setback, and the conflict had demonstrated that Denmark's status as a small state not only provided opportunities to the company but also set limits on its operations and expansion. Still, the conflict revealed the remarkably strong position that the GNTC already held, despite its size, for the Eastern was not able to crush the Danish company and win the war.[29]

The two companies signed an agreement of equality whereby they would in future share the GNTC's monopoly in China and act in concert toward the Chinese government. Eastern had in the meantime laid its own cable between Hong Kong and Shanghai in 1883, and the two companies now divided the region. Eastern promised not to expand north of Shanghai, and the GNTC promised not to move south of Hong Kong. Furthermore, they established a joint purse and adopted a cartel-like approach. This gave them total control of telegram traffic in the Far East and the links to Europe.

The agreement was obviously settled to avoid competition, but from a business as well a technical approach it also ended the final consolidation of system building in the region, which had followed the pattern of privately managed LTS businesses in America. It was a development that characterized the American telegraph in the 1850s and 1860s and that had been repeated by American railroad companies in the 1880s and 1890s, with the making of alliances and cartels in the process of building the continental railroad system.[30]

In America the system building in telegraphy was followed by mergers, but this did not happen in the case of the GNTC and Eastern and Associates despite the—at least on paper—unequal partnership. With a share capital of £1.5 million, the GNTC was by far the largest stock company in Scandinavia, but compared to Pender's telegraph "empire" it was a dwarf. The Eastern and the Eastern Extension had capitals of £3 million and £3.8 million, respectively, on top of which came Pender's other companies, including the powerful Telegraph Construction and Maintenance Company. Measured in terms of the size of the two cable networks, the Danish company also was small. While in 1892 the Pender group owned 112,711 kilometers of submarine cables, equal to 45.5 percent of the overall length of cables in the world, the GNTC accounted for only 12,838 kilometers, or 5.3 percent.[31]

From a business point of view the GNTC had a competitive advantage because the trans-Siberian link was both shorter and faster than Eastern's cable connection to Europe. The economic strength of the Pender group, however, was so overwhelming that a British acquisition of the GNTC would have been a "natural" and predictable result of the conflict if international telegraphy had been an "ordinary" business field. This, however, was precisely not the case because of the dominant political imperative. The Danish advantage was based on the GNTC's monopolistic concessions in Russia, which neither Eastern nor any other company could simply take over. The same could be said about the Japanese concession, which had become monopolistic in 1882.

The struggle therefore resulted in regional system building, with the GNTC's Far Eastern network filling the gap between China, Japan, and Russia in the same manner as its Scandinavian network already had in northern Europe. At the same time, the trans-Siberian link duplicated the British route between Europe and the Far East, a redundancy that benefited both companies, since they had agreed to have their two networks back each other up in the case of a breakdown or an overload of traffic. In all, the agreement established the foundation of a strategic alliance between the GNTC and the Eastern and Associated Telegraph Companies, in which the Danish company's network, from a British point of view, became an extension of the "nerves of the British Empire" to destinations outside the empire—mainly China, Japan, and Russia, but also the Scandinavian countries.[32]

The alliance was further strengthened when the two companies acquired a majority holding in and thus control of the U.S.-based Commercial Pacific Cable Company, which in 1903 laid the first and only American telegraph cable across the Pacific Ocean to the American territorial possessions of Hawaii, Guam, Midway, and the Philippines. In 1906 and 1907 the company extended its links to China and Japan, establishing telegraphic connection between those two countries and the American continent. The initiative to this new common action had been taken by the GNTC back in 1896 to avoid an independent American company laying and operating a cable across the Pacific. The Eastern Companies, which for years had tried to prevent the laying of a British, public-owned cable between Australia and Canada, were aware of the new threat and agreed.[33]

A major problem, however, was to overcome American legislative and political conditions. According to these, the company to lay the Pacific cable should be entirely American, without any foreign interests. Furthermore, it would not be allowed to enter into agreements regulating tariffs or sharing profits. In the words of the GNTC managing director, Commodore Edouard Suenson, the task therefore was to find an American partner who would help by establishing "a straw company for European interests."[34]

In 1901 John W. Mackay, president of the Commercial Cable Company and the Postal Telegraph Cable Company, agreed to play the role. Together with the Europeans, he formed the Commercial Pacific Cable Company with 50 percent of the capital of $3 million belonging to the Pender group, 25 percent to the GNTC, and the remaining 25 percent to the Mackay companies. In an additional agreement, the European shareholders committed themselves to act in unity within the American company and to maintain the internal distribution of shares.

The result was another step in the system building of the Far Eastern telegraphs, which was formalized on June 26, 1904, when the two former competitors together with a few other smaller telegraph companies and the Chinese and Russian telegraph administrations established a Pacific Joint

Purse and signed 17 secret agreements regulating telegram traffic and setting tariffs for the Pacific region and the links to Europe and the United States.[35]

By this arrangement, the GNTC and the Eastern Companies gained almost total control of all international telegraph links throughout East Asia and the Pacific—with the British partner as far the larger. It would hardly be fair, however, to define the Danish company as being "nominally Danish but secretly part of the Eastern cable empire."[36] Through the process of system building, the networks and services of the two companies had become integrated in the Far East and the link to western Europe. But, from both an economic as well as a political point of view, the GNTC had not been reduced to being simply a minor part of the British cable empire. Rather, the GNTC was already in a situation where—to use Schønebeck's phrase of 45 years later—its "absolute loyalty toward all the governments with which it collaborates" was essential to its survival. In the current situation, however, a special consideration of the interests of the Eastern and Associated Telegraph Companies in the Far East became an integrated part of the GNTC's strategy and decision making, thus further complicating the existing need to balance the interests of the great powers.

A Change of Conditions

The first years of the GNTC's existence had been a period of growth and remarkable expansion into the Far East and the Pacific rim: the company vigorously exploited its status as an independent, private enterprise from a neutral and small state. But the opportunities that had emerged for GNTC had done so within the framework of the age of empire, and when the framework started to change, so did the business environment and the political imperative in international submarine cable telegraphy.

Following the Spanish-American War of 1898 and the Boer War of 1899–1902, the great powers, as pointed out by Daniel Headrick, began increasingly to focus on the foreign policy and national security aspects of international telecommunications links: "As the reliance of nations on their communications increased, so did their fear of losing them." In fact, the GNTC had already felt the consequences of this "politicizing" of international telegraphy, as Headrick characterizes the new development, when it took the initiative to lay the Pacific cable from America to the Far East.[37] In this case, Denmark's political status was not sufficient to meet the demands from the American government, and instead the Commercial Pacific Cable Company was founded. On the whole, politicizing was a development that set new limitations on the expansion of the GNTC. Besides the laying of a cable in 1906 from Scotland to the Danish possessions of the Faroe Islands and Iceland in the North Atlantic, the company did not establish any new cable links before World War II. Politicizing, however, was also a development that led to threats against its existing cable network—threats that became still more serious as the process of politicizing was followed by profound changes in the existing world order.

In the beginning of the twentieth century, the age of empire began to transform into what Eric Hobsbawm has characterized as the "Age of Extremes"; the coming of communism, fascism, Nazism, and dictatorship resulted in profound changes in world politics and the relations between the great powers as well as fundamental new relations between business and politics on both national and international levels. To the GNTC the development was felt very heavily, as the company's status as a neutral company from a neutral small state began to lose its significance. The possibilities for political maneuvering among the great powers were constantly narrowing and, at the same time, the company experienced much more restrictive government policies in relation to international telegraphy and the flow of information that passed through the cables.

The GNTC had to operate in constant uncertainty, as it was not only operating in a strategic, high-risk sector but also dealing directly with governments and was thus totally dependent on concessions and landing licenses in a number of countries. The company had to confront the new situation where the management of the political imperative in international submarine telegraphy had increasingly become the management of political risk—all the more since the development most profoundly affected two countries of crucial importance to the GNTC: Russia and Japan.

Ring, Lenway, and Govekar deal with host country political risks as well as home country political risks. With the coming of the "Age of Extremes," the GNTC faced increasing and dangerous host country risks, but it was never subject to hostile political interventions in its home country. On the contrary, the GNTC could continue to rely on the unconditional support of the Danish state. In 1929 the Danish minister of foreign affairs openly declared in parliament that "the GNCT is in fact a sub-department of the Foreign Ministry and when we have this strong diplomatic representation in the Far East, it is to a large degree exactly in order to safeguard the GNTC's interests out there."[38]

The Danish Ministry of Finance had sold its shares in 1882 when the GNTC was stabilized after a financial crisis in the late 1870s, but the company nonetheless continued to be fully backed by the Danish state, no matter which political party formed the government. Prominent politicians and ministers sat on the company's board of directors, including the Danish prime minister from 1920 to 1924, Niels Neergaard. Denmark's diplomatic influence was, however, rather limited, and the GNTC often preferred to act on its own—and not always in accordance with official Danish foreign policy. In many ways, the GNTC often performed its own diplomacy, and this required foreign policy and diplomatic skills in the management. From the very beginning, senior diplomats or persons with other relevant qualifications, such as international law, were elected to the board, and in the most important countries of Russia and Great Britain the company set up its own "diplomatic" representations to deal with governments and authorities. In the Far East, the representative was situated in Shanghai and held the status of general manager dealing with the Chinese and Japanese authorities as well as with the foreign diplomats and the international settlements.

New managers were handpicked, sent abroad, and trained in dealing with foreign authorities and cultures, or they were recruited from the Ministry of Foreign Affairs or other employment where they had already obtained political and diplomatic experience. An example is the Danish senior diplomat Otto Scavenius, who was elected to the board in 1919 when he was director of the Ministry of Foreign Affairs. For a short period in 1920 he was actually minister of foreign affairs and at the same time a member of the GNTC's board of directors. When he left the Ministry in 1924, he was appointed managing director of the GNTC, and in 1938–1945 he was chairman of the board.[39]

All in all, the GNTC in several important aspects was both organized as and behaved as a political or diplomatic organization. There is little doubt that the company's managerial expertise helped it to handle the extremely complex situation and serious political threats it faced in its relations with Russia and Japan in the first half of the twentieth century.

Mutual Dependency and Interests

In their framework for analyzing the political imperative in international business, Ring, Lenway, and Govekar introduce the concept of strategic interdependence, defined as "a relationship in which the state and the MNC [multinational corporation] have to deal with each other to accomplish their objectives." In their work, the concept is related to ownership/control and MNCs' ability to transfer risks by either forestalling or absorbing them in order to maintain their independence—depending on their status as either global or multidomestic industries.[40]

Again, these concepts are not directly applicable to international telegraphy because of the LTS character of the industry. Yet international telegraphy was based on strategic interdependence, since companies and governments had to deal with one another in order to create international cable links. This was especially the case with the GNTC, whose telegraph system depended totally on cooperation with a number of countries, including Russia and Japan, which on their side benefited from the establishment of the Danish telegraph network.

With the laying of the cables in the Far East and the opening of the trans-Siberian link, the GNTC had achieved a position of crucial importance to Russia. Not only was the country connected to the global telegraph network through the company's cables, but the Russian state—as well as the Danish—benefited economically in the form of large transit revenues from the telegram traffic between Europe and the Far East. Thus GNTC could always reckon on Russian diplomatic support, and in 1882 the GNTC's Russian concessions were prolonged until the end of 1912.[41]

Japan also benefited from its cooperation with the Danish company. To the Japanese leading circles, which after the Meiji revolution were eager to adopt Western technology, the Danish company seemed an ideal partner since Japan had neither the financial nor the technological capabilities to lay its own cables. Therefore the Japanese government granted the company a 30-year concession, including a nonmonopolistic license to land its cables at Nagasaki and erect its own telegraph station to operate the cables. In December 1882, the concession came to include a Danish 20-year monopoly on Japan's international telegraphy as a "payment" for laying a cable between Japan and Korea. In March 1900 the concession, including the monopoly, was extended until the end of 1912.[42]

To the GNTC, the cooperation with Russia was essential, not because of Russia's own rather limited international telecommunications, but because of the trans-Siberian link that connected Europe and the Far East. The structure of the GNTC telegraph system could be divided into the three geographical regions where the company ran its main business: northern Europe, the Far East and the trans-Siberian link. It was, however, the trans-Siberian link that constituted the foundation of the company's strength and position in international telegraphy by connecting its two cable systems in Europe and the Far East. Without the trans-Siberian link, the company's system would be reduced to two regional cable networks, thus diminishing the importance of the company in the global business of international telegraphy. The GNTC, therefore, depended on good relations with the Russian government.

In other words, the two partners were tied together by common interests and—most important—mutual dependency that despite differences and controversies formed the basis of a fundamentally good relationship.[43]

The same could be said regarding the GNTC's relations with Japan. The company's main interest was not Japan's own international telecommunications, but the use of Nagasaki's position as a stepping stone for the cables between Shanghai and Vladivostok. For this reason, the concession, with its landing license as well as the right to operate the cables from the Nagasaki station, was of great strategic importance to the company, comparable with the importance of the trans-Siberian link.

Japan was just as dependent on the GNTC, since the Danish cables were the country's only telecommunications links to the outer world until 1906. Furthermore, the country desperately needed connection to the global telegraphy network in its modernization process, so the cooperation with the GNTC was of great strategic importance. Thus, as was the case with Russia, the two partners were tied together by mutual dependency and interests, and for this reason the relationship was fundamentally solid, at least in the beginning.[44]

Japanese Expansionism

By the turn of the century, a change occurred in the relations between the GNTC and Japan. In contrast to Russia, Japan had developed the technological capacity to lay and operate its own submarine telegraph cables—a status hitherto reserved to a few Western countries—and demonstrated this by constructing a cable link to Taiwan in 1897–1898. The laying of the cable was a result of the annexation of Taiwan after the Sino-Japanese War of 1894–1895, an act that transformed Japan into a colonial empire. As the incident became a turning point in Japan's modern history, it also became a turning point in the country's technological development.[45]

Japan's geographical expansionism was followed by telegraphic expansionism, propelled by the desire to create an imperialistic telecommunications "nerve system" and to gain control over the country's international telecommunications. This brought the country onto a collision course with the GNTC, and it was only because the duration of the concession was tied to the Chinese and Russian concessions, which both were extended until the end of 1912, that the Japanese government in March 1900 accepted a similar extension of the GNTC's concession.[46]

The emerging Japanese consciousness and nationalism in telegraphic matters coincided with the generally growing awareness in world politics of the importance of telecommunications. With the Russo-Japanese War of 1904–1905, this connection became both acute and evident to the two countries, as well as to the GNTC. Japan was concerned about the GNTC's close relations with the Russian government and, even though the company remained strictly neutral in the conflict, the Japanese navy cut the cables between Nagasaki and Vladivostok. They were repaired after the peace settlement, but the conditions for the GNTC's business in the Far East had radically changed.

As far as Russia was concerned, the GNTC's relations became further consolidated in 1906 with a 20-year extension of its monopolistic concessions as part of an agreement on laying a cable between St. Petersburg, Libau, and Denmark. The Japanese government, however, challenged the GNTC when it opened, in the same year, a military cable to the Asian mainland for public use. The GNTC protested, claiming that Japan was violating its monopolistic concession, but its objection was rejected. Instead, the Japanese government invited the company to send a representative to Tokyo for preliminary negotiations over the concession—even though this would not expire until the end of 1912.

Undoubtedly, the Japanese action was part of the government's policy of treaty revision that was aimed at abolishing the extraterritorial rights of foreign countries and companies, which included the GNTC telegraph station in Nagasaki. But it was more than that: Japan wanted to gain control over its international telecommunications and to pave the way for total independence from foreign influence, thus ignoring the GNTC's status as a company from a small, neutral state. First and foremost, the government demanded a termination of the GNTC's monopoly. The company had, in fact, already decided that it was prepared to relinquish this. However, when it came to the Japanese demand that the government should be allowed to lay its own cables to Russia and China and use them for all kinds of telegrams, including transit traffic to and from the United States, the GNTC refused. The company was fully entitled to do so because China, in August 1900, had extended its monopoly until the end of 1930. If Japan wanted to land a cable on the Chinese coast, the GNTC, according to the terms of its Chinese concession, should give its consent—which it did not.[47]

Referring to its concession from 1882, according to which the GNTC "in any case" should enjoy the right to operate its cables in connection with the Japanese domestic telegraphs, the company rejected that its rights to land and operate its cables at Nagasaki should be modified and limited in time. The Japanese government pointed out that the word "permanent" had no place in legal documents, but the GNTC stood its ground, claiming that its rights should be valid indefinitely and permanently.

The negotiations were crucial to both Japan and the GNTC. Nonetheless, a really serious crisis never developed since the company was well protected by its Chinese and Russian concessions. On the whole, despite its growing military and political strength, Japan fully accepted the GNTC's concessions as well as the International Telegraph Convention as the basis for entering a new agreement with the company. The Japanese government continued to question the GNTC's vindication of the permanence of its landing rights but avoided bringing matters to a head—most probably because the Western powers would certainly have sided with the GNTC because of the strategic importance of the trans-Siberian telegraph link.

Nevertheless, it was only after six years of negotiations and diplomatic assistance and pressure from Great Britain and several other countries that an agreement between the GNTC and Japan was signed on August 23, 1913. Most important, the GNTC's right to continue to land and operate its cables in Japan was confirmed, with no fixed time limit. Japan was allowed to lay a cable to Shanghai on the condition that it was to be used exclusively for terminal traffic in Japanese characters and for Japanese-Chinese government telegrams. Furthermore, the Japanese government had to accept that a joint purse should be established for terminal traffic between China and Japan, with 64.5 percent of the total revenues from both the Danish and Japanese cables going to the GNTC.[48]

The company had successfully defended its business in Japan and, at the same time, secured the strategic position of its Nagasaki station. Its Chinese concession allowed the company to restrain Japan's telegraphic ambitions and to impede its rise as a dominant power within Far Eastern telegraphy. Nevertheless, the company, as its manager, Edouard Suenson, expressed during the negotiations, continued to hope that its "neutrality" and general standing would still make it an attractive partner to Japan.[49]

On the contrary, however, the negotiations resulted in a serious worsening of the relations between the two parties. Japan's position was hardened by the GNTC's obstinate standing on its concessional rights and its unwillingness to comply with the Japanese demands that left no room for future compromises. Instead, the conclusion of the new agreement came to mark the beginning of a long and fierce conflict, with Japan in the offensive role, constantly seeking to undermine the company's position, especially in China. This served to prepare the ground for the final showdown in 1930, when the GNTC's Chinese monopoly would expire.

Russian Revolution and Bolshevik Power

For the GNTC, the continuing controversies with Japan were totally overshadowed when the czarist regime was overthrown in 1917 and Lenin's Bolsheviks took control of Russia. The Soviet government immediately announced that it refused to recognize the GNTC's concessions from the czarist period, and, in contrast to the Japanese government, it did not have any respect at all for the International Telegraph Convention—or any other international conventions and agreements.

The company found itself in a very unsafe and unpleasant situation that, with the outbreak of the Russian Civil War in the summer of 1918, developed into its worst crisis yet. Military operations soon resulted in the physical interruption of the trans-Siberian link, and adding to this came the political and diplomatic complications when the Western powers—and Japan—sided with the Whites and implemented an all-embracing blockade of Soviet Russia. Denmark attempted to stay neutral but was forced by Britain to join the blockade, which expressly included post and telegraph.

The GNTC shared the general dislike of the Bolsheviks, but business interests forced it into a political-diplomatic double game. The company, avoiding any public comment on the Bolsheviks, attempted to maintain links to the Soviet government, even after Western diplomats, as well as the company's employees and its representative, had been officially recalled from Soviet Russia. At the same time it supplied the Western powers with telecommunications links to Admiral Koltjak and his government as well as to the Allied High Commission in Omsk.[50]

Good relations with Russia, irrespective of whether it was ruled by the czar, the Bolsheviks, or admiral Koltjak, were of crucial importance to the GNTC. Therefore, it decided to back both horses until the Civil War produced a clear winner. In late March 1920, the company broke the international blockade and took the initiative to enter secret negotiations in Copenhagen with the Soviet deputy foreign commissar, Maxim Litvinov.

The leading Russian Bolshevik was more than positive, since the Soviet government urgently needed to restore the country's international telecommunications links in both Europe and the Far East, and so within a week an interim agreement was signed. As the reopening of the trans-Siberian link was also in the interests of Great Britain, the Foreign Office soon gave the go-ahead. Thus the way was paved for the GNTC to gain a new Russian concession. In its recommendation to the Soviet government, the Commissariat of Foreign Affairs emphasized that the country needed "reliable telegraph links with the outside world" and that the GNTC—in contrast to its British competitors—was "politically, totally independent" and only interested in "commercial profit."[51]

It could be said that the Soviet government had little choice insofar as the GNTC owned most of the cables connecting Russia to the outer world—a situation that the country shared with Japan. The archives, however, leave little doubt that the company's political neutrality made it an ideal partner for the Bolsheviks in international telecommunications (despite Denmark's participation in the blockade).[52]

In the Kremlin on July 21, 1921, Lenin signed the concession that would run until the end of 1946. Not only was it the first concession to be granted to a foreign company after the revolution, but the importance to the Soviet government of the reopening of the telegraph links to Europe and the Far East is apparent from the fact that in several respects the concession was more advantageous to the GNTC than those granted by the czarist government. Furthermore, the Bolsheviks went against their own principle not to grant a concession to a company whose government had refused to enter diplomatic relations with Soviet Russia. (Denmark did so only in 1924.) Quite remarkably, they also agreed to pay the company an outstanding debt of 7 million gold francs, a debt originating from the czarist telegraph administration.[53]

In January 1922 the GNTC was able to reopen its telegraph cables to Russia and on March 25, 1922, the trans-Siberian link was reopened after almost four years. It was operated by more than one hundred Danish and equally many Russian telegraph operators on GNTC stations throughout Soviet Russia, from Petrograd to Vladivostok. An official GNTC representative, Alfred Schønebeck, was stationed in Moscow with his own office and residence.

Controversies in Soviet Russia

During the first few years the collaboration between the GNTC and the Soviet government was friendly and without any major controversy. In the words of Alfred Schønebeck, the company benefited from its status as coming from "a small and politically harmless country." But in the late 1920s the situation started to change.[54]

In spring 1926 the Soviet authorities announced that, in their opinion, their debt had been fully repaid. When the GNTC protested, claiming that they still had 2 million gold francs to their credit, the People's Commissariat of Post and Telegraphy simply threatened to terminate the company's concession. Such a step would represent a violation of the International Telegraph

Convention. At the same time Schønebeck received a confidential communication that the Soviet Supreme Concession Committee was considering changing or even terminating the concession, so he knew that the threat was serious.

Nevertheless, the company stood firm on its position and appealed to the Supreme Concession Committee, which in 1928 settled the dispute in favor of the People's Commissariat. The decision came as no surprise to the GNTC, which doubted the impartiality of the Soviet legal institutions. For the same reason, they gave up the idea of taking the matter to court in Moscow. The GNTC kept its concession and even succeeded in including an additional clause that allowed it to duplicate the trans-Siberian landline, but the controversy nevertheless came to mark a change in relations between the company and the Soviet authorities.[55]

Since 1926 Schønebeck had felt an increasing "tendency to suspicion and jealousy towards the company."[56] It was an ideological and nationalistic attitude, rooted in the rising stratum of Soviet specialists and bureaucrats who increasingly influenced the government institutions and administration. It was caused partly by the GNTC's being a foreign company and partly by the Soviet telegraph administration's desire to take over the company's operations. Undoubtedly these personal experiences, combined with the overall development in the Soviet Union of the late 1920s, encouraged Schønebeck to appear before Gosplan on January 31, 1930.

Apparently his action had the desired effect. The GNTC was not exposed to the pressure that was put on other foreign companies to make them leave the country. On the contrary, the Soviet authorities in February 1933 created a special "Lex GNTC" excepting the company from a new law according to which Russian-born women married to foreigners working in the country were not allowed to return to the Soviet Union if they traveled abroad—even if only for vacation. In May the same year, the managing director of the GNTC, Alf Ussing, paid a visit to Moscow, where he was received by the Post and Telegraph Commissar, Aleksei Rykoff, and a number of high-ranking officials. On his return to Denmark, Ussing reported that he had been treated "with the outmost friendliness" and that the Soviet authorities apparently had a "favorable" view of the GNTC. Everything seemed peaceful and quiet, without any immediate threat to the company and its concession.[57]

New Agreement with Japan

A striking difference between the developments in Japan and the Soviet Union was that Schønebeck was given access to one of the most important

Soviet institutions—and it was worth the effort. In Japan the GNTC had given up trying to reach an understanding about the situation after 1930, as the Japanese government had made it clear that there was nothing to negotiate. When the company that same year learned that Japan and China had entered an agreement to take over in 1930 the GNTC's cables between Nagasaki and Shanghai, the management realized that the company was facing what it had feared the most: a Japanese-led Sino-Japanese alliance against its position in the Far East. Japan obviously aimed at both finishing off the GNTC's position in the region and taking over the company's cable network herself. As stated by the managing director, the GNTC could only hope that "before 1930 there would come such a change in the political situation in the Far East that it would be impossible for Japan to carry out its intentions."[58]

The GNTC got what it was hoping for. Following the military clashes in May 1928 between Japanese forces and the troops of Kuomintang in Manchuria and the subsequent murder of a Chinese governor by Japanese officers, relations between Japan and China became extremely tense. As a result, the Chinese government gave up its telegraph alliance with Japan and, instead, entered into negotiations with the GNTC, as well as the Eastern and Commercial Pacific, which also operated in China. A preliminary agreement was reached at the very last moment on December 31, 1930, but a final settlement was delayed by the Manchurian Incident in September 1931 and the Japanese military operations around Shanghai in February 1932.

A conclusive agreement with expiration on December 31, 1944, was signed by China and the telegraph companies in April 1933, leaving no opportunity for Japan in this period to take over the GNTC's cables. And even though the GNTC now had given up its Chinese monopoly, Japan was still unable to lay its own cables to China because of the tense relations between the two countries. Instead, Japan had no choice but to enter an agreement herself with the GNTC, which was done a few months later on the basis of the 1912 concession. Following the new agreement, the Japanese government immediately opened up its existing cable to Shanghai for all kinds of traffic. At the same time it terminated the joint purse agreement with the GNTC, thus "getting free from one of the fetters that had been impeding the autonomy of communications in Japan," as Kaoru Hanaoka, a then employee in the Japanese telegraph administration, expressed it.[59]

When it came to the critical issue of the GNTC's right to land its cables and operate them from the Nagasaki station, however, Japan had to accept its permanence. Despite its strong dislike of the presence of a foreign telegraph station on Japanese territory and its general animosity toward the GNTC, the government continued to accept the company's interpretation of the 1882 concession. Even though the new Chinese and Japanese concession would mean economic losses for the GNTC, the company's strategic position in the Far East had been secured—at least until the end of 1944, when the Chinese concession would expire.

Stalinist Repression

In the mid-1930s the GNTC's relations to both the Soviet Union and Japan underwent profound changes. In Russia the change came suddenly and unexpectedly with the assassination of Kirov in Leningrad on December 1, 1934. The GNTC's station in Leningrad was immediately put under close surveillance by the Soviet secret police, NKVD (Narodnyi Komissariat Vnutrennikh Del), and several of its Russian telegraph operators were arrested. In early January 1935, the company was informed that seven of its leading Danish employees in Leningrad were considered "undesirables" in the Soviet Union and therefore had to leave the country immediately. No explanation was given, and even though the GNTC considered the expulsion to be a "severe abuse" of a clause in the concession giving the Soviet authorities no right to expel individual Danish employees without any warning or explanation, the company, after appealing to both Rykoff and to Litvinov (now commissar of foreign affairs), had little choice but to accept the Russian position.[60]

The action in Leningrad was accompanied by a demand from the Commissariat that fourteen of the GNTC's Danish employees in the Soviet Union should be replaced by Soviet telegraph operators. The company accepted, stipulating that the Soviet operators should be properly trained, and, after an agreement had been signed in May 1935, a year of a more peaceful and relaxed relationship followed, until the Commissariat in May 1936 put forward new claims. This time, the GNTC was asked to give up its station in Moscow and hand it over to the Russian telegraph administration—a demand that the company opposed for obvious reasons. However, it never came to full-blown negotiations because the Commissariat, to the company's wonder and great relief, suddenly fell completely silent on the matter. But this was only the lull before the storm.

On July 21, 1937, the GNTC representative in Moscow was called to the Commissariat, where he was informed that the Soviet government "for various reasons" no longer would allow foreigners to work at telegraph stations inside the Soviet Union. The government therefore demanded that the concession from 1921 be replaced with a new agreement that would ensure the company its usual profit but that would hand over the operation of the trans-Siberian landline to the Soviet telegraph administration, employing only Soviet citizens. It was emphasized that the Soviet leadership considered the matter to be "serious" and thus expected it to be resolved "very quickly."[61]

Without doubt, the reason for the Soviet demand had been indicated a few weeks before in an article on "some methods and tricks of the foreign espionage organizations and their Trotskyite-Bukharinist agents," printed in the *Leningrad Pravda* on June 11, 1937. In the article, written by the local head of the NKVD and one of Stalin's high-ranking henchmen, Leonid Zakovsky, the GNTC was mentioned in connection with German espionage in Russia during World War I. According to Zakovsky, German spies, "through a long period and with the help of the Danish telegraph company," had succeeded in sending secret information from St. Petersburg to the General Staff in Berlin. Zakovsky did not directly accuse the GNTC of espionage; neither did he refer to their present activities. The message, however, was clear: the Danish telegraph company was a threat to the Soviet Union's national security.[62]

Regardless of the reasons, the government's action against the GNTC was a violation of the concession that would expire at the end of 1946. But, after consulting the Danish Ministry of Foreign Affairs and several Danish experts on Soviet affairs, the GNTC realized that they had little choice other than to accept. The Soviet government obviously had made a definitive decision that could not be appealed, and with its firsthand knowledge of the Stalinist regime the company feared the consequences of standing up for their concessional rights and rejecting the Soviet demand.

This did not mean, however, that the GNTC gave up its concession completely. First and foremost, the company feared that a renegotiation of the concession might lead to new Soviet demands and maybe even to the closing of the trans-Siberian link. Under any circumstances, the company would have to inform its partners, both governments and companies, that it had entered a new agreement with the Soviet Union. As a result, a number of countries, including Japan, could demand a renegotiation of their agreements with the GNTC. This was also the case with Cable and Wireless (the former Eastern and Associated Telegraph Companies) and the Commercial Pacific Cable Company. The GNTC also feared the reactions among its customers if it became known that telegrams passing through Russia would be handled by Soviet citizens rather than by Danish telegraph operators.

Despite Soviet assurances that the GNTC would continue to receive its usual profit, the conclusion of a new agreement would most probably result in serious damage to the company and its position. With a focus on the

Soviet government's major concern, the presence of Danish citizens at the telegraph stations, the company therefore suggested an arrangement whereby all its Danish employees should leave Russia and hand over the operation of the telegraph link to the Soviet telegraph administration and its personnel. The GNTC should still be allowed to have an official office as well as a permanent representative in Moscow. The arrangement should be settled through an amendment to the existing concession that should continue to be valid officially—and all this should be kept secret to avoid unwanted and unnecessary public attention.[63]

The company explained its proposal by pointing out the considerable economic losses that the Soviet Union would suffer if the traffic on the trans-Siberian link should decline or even disappear. In 1935 the Soviet state had received 2,163,234 hard-currency rubles in transit revenues and in 1936 the amount was 2,134,318. On this basis the company's suggestion was supported by the Commissariat of Post and Telegraphy and the Commissariat of Foreign Affairs. But since the whole matter was the result of a decision in the Soviet government to "liquidate" the GNTC's concession, the proposal had to be laid before Stalin personally. This happened on September 8, 1937, and, with the supreme leader, Joseph Stalin's, acceptance, the Commissariat now was able to follow the company's suggestion.[64]

By the end of March 1938, all the GNTC's Danish employees had left the Soviet Union except for the representative in Moscow and two telegraph operators who had been arrested by the NKVD on allegations of espionage. The arrests had nothing to do with the GNTC's activities and position in the Soviet Union; they were simply an incidental result of the purges that harried the country. As a gesture to the company, both were sentenced to expulsion from the Soviet Union and in February 1939 they were back in Denmark, marked by the interrogations and the inhumane conditions in the NKVD prisons but alive.[65]

Despite Stalinism and the regime's overall hostility toward and suspicion of foreigners, the GNTC still enjoyed a favored position in the Soviet Union. In 1921 it had been the first foreign company to be granted a Soviet concession, and now it was the only company that did not have its concession withdrawn in the 1930s. This was all the more remarkable as the Soviet government in 1937 had abolished the Supreme Concession Committee following a government decision to eliminate the concession sector in the Soviet economy.[66]

For some reason, however, the amendment to the company's concession was never formulated or signed. The Soviet government, in August 1939, decided to raise the matter anew, but because of the outbreak of World War II in September 1939, and the German occupation of Denmark on April

9, 1940, it was never presented to the company's representative in Moscow. Following the German attack on the Soviet Union on June 22, 1941, it no longer mattered to either the Soviet government or the GNTC.

A Foreign Enclave within Japanese Territory

Parallel to the difficulties in the Soviet Union, the GNTC was also facing increasing pressure in the Far East, following the transition of Japan into total dictatorship. The controversies began to accelerate in 1934, when the Japanese telegraph company in the puppet state of Manchukuo, MTT, started to disregard the customers' "via" indications, for example, "via Northern" or "via Eastern." The companies protested, apparently to no avail, pointing out that it was a violation of the International Telegraph Convention.[67]

Relations grew more tense during 1935 as the Japanese telegraph authorities launched a campaign trying to attract the GNTC's customers to their own wireless service. It was a campaign in accordance with the official "Encourage Japanese Manufacturing and Buy Japanese" movement, but nonetheless a rather aggressive move with a clear target. In their enthusiasm, local branches even sought out some of the Danish company's major commercial customers and indicated that they might receive poorer service at the local telegraph offices if they continued to send their international telegrams via Northern. The GNTC protested, but again with little success. Instead, the Japanese telegraph administration struck back, accusing the company of having contacted customers directly, in violation of the 1912 concession.[68]

It was, of course, much more than a matter of fierce competition. As part of Japan's policy of reducing hard currency payments in order to ensure the import of strategic and vital resources, the wireless campaign aimed at decreasing the GNTC's income from its international cable service between Japan and the rest of the world. In 1935 over ¥10 million were paid to the company, while Japan, unlike Russia, had no income at all from the GNTC's transit traffic since the telegrams did not pass through Japanese territory.[69]

Most important, however, was Japan's desire to extend its influence throughout Asia. It made complete autonomy in international communications a strategic priority. As pointed out by Kaoru Hanaoka, "The GNTC's station constituted a foreign enclave within the territory of Japan. . . . This foreign enclave within the territory of Japan was a matter of primary concern as well as a serious and difficult problem for the telecommunications authorities of Japan in the Meiji and Taisho areas. It was also true that, each time a national emergency arose, the national defense and the telecommunications authorities of this country took extraordinary precautions against this enclave."[70]

In this context, a climax was reached with the government's adoption, in August 1936, of the "Fundamentals of National Policy," proclaiming Japan to be the stabilizing power in Asia as the basis of a new imperial strategy directed against British and American influence. As reflected in the Japanese press and the government-controlled propaganda, the government was worried about the GNTC's close relationship with Great Britain. The company was referred to as a "British" or "British-controlled" company and was accused of being an "agent of British influence."[71] In early February 1937, the Japanese Ministry of Communications began to address the GNTC issue in a new and fundamentally different manner, considering different ways to eliminate the company's presence in Japan and Far East.[72]

Ending the GNTC in Japan

Japan's full-scale military attack on China in July 1937 was soon followed by Japanese actions against the GNTC and the two other telegraph companies whose offices and stations were situated in the GNTC's building in Shanghai. In November 1937 the Japanese military authorities in Shanghai insisted on Japanese military censorship of international telegrams. They also demanded that the companies no longer pay terminal fees to the Chinese government. The companies protested, and so did the consuls-general from Denmark, Great Britain, and the United States as well as the Chinese government, but without result.[73]

Instead, the Japanese military authorities in July 1938 announced their intention to take over the Chinese telegraph offices that had been set up inside the GNTC's building in 1933 to handle international telegrams sent by the three companies. Again the three companies and their consuls-general protested, again without success. Instead, the companies accepted the Japanese demands "due to force majeure."[74] After this, the three companies were held in an iron grip by the Japanese military authorities, which now had their personnel stationed inside the GNTC's building.

Parallel to the developments in Shanghai, ministries and authorities in Tokyo had examined various scenarios for the final elimination of the GNTC's presence in the country. There was a strong and real suspicion of espionage that, however, proved to be difficult to substantiate, despite constant surveillance of the company and its employees. Nevertheless, voices in the Ministry of Communications were raised in support of the use of coercive measures. But in February 1939 personnel in the ministry rediscovered the technical clause in the concession stating that "the Company's station in Nagasaki shall not be allowed to have direct dealing with the public."[75]

On this basis, the ministry decided to accuse the GNTC of illegal business methods and to revoke its concession immediately. But before confronting the company, the ministry had to obtain full agreement from other ministries and authorities. Everybody was in favor of ending the GNTC's presence in Japan, but when it came to the methods to achieve this goal, the differences in opinion were considerable. The Ministry of Communications was supported by the army and the navy as well as the Home Ministries, while the Foreign Ministry was much more cautious. It was not until early March 1940 that a compromise was reached. The GNTC's concession was to be revoked by the end of April 1943, thus giving the company three more years of business, but with the restriction that the cables should be operated by Japanese personnel.[76]

On March 18, 1940, the GNTC's general manager in the Far East was informed in Tokyo about the Japanese decision. It was emphasized that the decision was "final and without possibility of compromise," and the company was also warned against demanding evidence for the allegations of violations of the concession, since this could have "serious consequences" for both individuals and the GNTC. Between the lines it was hinted that some of the company's employees had been involved in espionage.[77]

The government demanded an answer—or rather an acceptance—from the GNTC within four weeks. The general manager left for Shanghai, from where he consulted the Danish Ministry of Foreign Affairs and the GNTC headquarters in Denmark. After two weeks he was convinced that there was no alternative to accepting the Japanese decision and on April 9, 1940, he returned to Tokyo to sign the new agreement. On the very same day Denmark was occupied by German military forces, disrupting all of the country's telecommunications links with the rest of the world. Even though the link to the Far East was reopened after a few weeks, the company's international telegraph system was broken apart. With the German invasion of Stalinist Russia on June 22, 1941, and the Japanese attack on Pearl Harbor on December 7, 1941, the GNTC's Japanese concession was just a piece of paper. When it finally expired on April 30, 1943, it was hardly noticed in the company's headquarters.

Conclusions

The definitive and decisive actions against the GNTC in both Soviet Russia and Japan were obviously caused by the two countries' preparations for war. An important distinction, however, was that Russia was preparing for defense, while Japan was preparing for attack. And while Russia was apparently seeking a new kind of arrangement with the company, Japan was quite openly aiming for the elimination of its presence in the country and in the whole region.

As happened in most other countries, with the United States as the exception, Russia and Japan from the very beginning built and operated the domestic telegraph networks themselves, leaving it to private companies to establish and operate their international, overseas telecommunications links.[78] In this connection, the GNTC had been an ideal partner for both Russia and Japan earlier on. But as the two countries and their national policies changed over the years, their views on international telecommunications, including their cooperation with the GNTC, changed as well.

While reflecting the different national policies of the two regimes—isolationism and expansionism, respectively—the actions against the GNTC nevertheless had a single important and identical purpose: to establish state management of the country's international telecommunications and thus to gain total control over the flow of information in and out of the country. Of course, this was a result of historical circumstances—increasing political tensions and a growing threat of war in Asia and Europe. Other powers in similar situations reacted with the imposition of temporary military censorship, as happened in Russia before the Revolution. With the worsening of the international relations following Hitler's rise to power and Japanese expansion in the Far East, Great Britain also took precautions to ensure the control of its international telecommunications links, but it never took action against the GNTC by trying to gain direct control over the Danish cables landed in Great Britain. In contrast, the arrangements in Soviet Russia and Japan were directly aimed at the Danish company and were intended to be permanent.

As pointed out by Daniel Headrick, the security aspects of international telecommunications per se were "political and organizational of nature."[79] To governments, it was equally important to obtain reliable and confidential information as it was to prevent other governments and nations from obtaining the same kind of information. Seen against this background, the GNTC's status in coming from a small, neutral state did not matter much to Stalinist Russia or to fascist Japan.

Japan's desire to control its own international telecommunications was deeply rooted in nationalism and was as old as the introduction of the telegraph in the country. As soon as Japan had developed the technological and financial capabilities to lay and operate submarine cables, the government began to seek autonomy from the GNTC. With the simultaneous transformation of Japan into a colonial power, the government recognized the importance of having an imperial "nerve system" of telecommunications links, and with the rise of Japanese militarism and fascism, it was only a matter of time before the strains on relations would come to a head. When the government decided, in August 1936, on Japan's new role and the subsequent military attack on China, the GNTC's Far Eastern cable network was situated "inside" the Japanese Empire, and in spring 1940 the GNTC's presence was eliminated from Japanese territory.

In Russia the development was quite different. From the very beginning czarist Russia had been a colonial empire, but without the overseas territories that demanded a "nerve system" like Japan's—or Great Britain's. Russia was fundamentally a continental power, and the establishment of international, overseas telecommunications links had been mainly for strategic purposes related to international politics and national security. The Russian Revolution did not change the Russian position, even though it affected relations with Denmark. The Bolsheviks apparently had confidence in the Great Northern Telegraph Company, and it was only after the development of full-scale Stalinism and mass repression that the regime took steps in the mid-1930s to eliminate the company's presence in the country.[80]

For both Japan and Soviet Russia it was obviously the GNTC's foreign status, and the presence of Danish citizens at its stations, that was the cause of their concern. In this respect, the Japanese view of the Nagasaki station as being a "foreign enclave" within the country's territory hardly differed from the Russian position. Nonetheless, the two regimes had different perspectives in their actions against the GNTC. Japan aimed at the total elimination of the company's business in Japan and the Far East, while Soviet Russia offered a continuation of the cooperation, based on a new agreement that would even ensure the GNTC its usual profit.

These different perspectives reflect and express the profound difference in national policies between imperialistic expansionism and Stalinist isolationism. While the GNTC was a hindrance to Japan's own telecommunications expansionism, this was not the case in Russia, which lacked such ambitions. Furthermore, the Soviet Union had nothing to gain by eliminating the GNTC as long as the trans-Siberian link was handled by the Soviet telegraph administration. On the contrary, the Soviet government had a financial interest in keeping the GNTC in business because of the transit revenues from the trans-Siberian link. Without doubt this was exactly the point that made Stalin change his opinion on the "liquidation" of the GNTC's concession.

When it came to enforcing the decisions to end the GNTC's concession, the two regimes also took remarkably different steps. While the Japanese government spent a long time in search of a formal pretext, the Soviet government simply informed the company of the government's decision. That Stalin did not have any respect for the written concession is not surprising, any more so than his reversal of the government's decision to liquidate the GNTC's concession. That Japan offered the GNTC a three-year extension of its landing license seems to have been a case of face saving—a gesture that had no room in Stalinist policy.[81]

The reasons, however, for Japan's having waited so many years before fulfilling its desire to get rid of the GNTC and for finding it necessary to legitimate the action by inventing a pretext in the wording of the concession are unclear and difficult to understand, especially if one takes into account the general character and the behavior of the Japanese regime. Most probably, the Japanese government feared international complications and, especially, the reaction from the Western great powers, which could be expected to side with the GNTC. Or maybe the government simply expected that the continuing pressure on the small company from the small, distant country would, sooner or later, force to it to surrender. If the latter was indeed the case, the Japanese government seriously miscalculated.

"The telegraph knows no politics" was the motto of the founder of the mighty Eastern and Associated Telegraph Companies, Sir John Pender. The history of the GNTC and its relations with Russia and Japan, however, tell another story. As indicated by its early development as well as its relations with Russia and Japan in the first half of the twentieth century, the GNTC based its business not on ordinary market conditions but on the management of the political imperative. It is a development that reveals how governments were concerned about the political and security aspects of the submarine cables as well as the flow of information they provided. At the same time, it provides valuable insight into the circumstances under which international telegraph companies had to do business before World War II.

The negative effects of the Japanese and the Soviet governments' intervention in international telecommunications in the 1930s were felt severely by the GNTC. The long-term historical development of the relations between these two countries and the GNTC, however, reveals that the political imperative is not only a matter of political risk but also of political opportunity. The founder of the GNTC, C. F. Tietgen, was fully aware of the political imperative of international telegraphy. In fact, his whole business strategy in founding the GNTC was built on political analyses in which he was perfectly conscious of both the opportunities and the risks for a company coming from a small, neutral state.

In Russia and Japan the regimes and national policies gradually changed in a way that radically altered business conditions for the GNTC.

The company's advantage of coming from a small, neutral country no longer counted, as the regimes' previous acceptance of foreign control over their international telecommunications became overshadowed by new and far more important political issues. The political imperative became political risk, leaving the GNTC with its back against the wall, fighting for survival.

In Japan the change came around the turn of the century. It actually was quite remarkable that the GNTC succeeded at all in withstanding the Japanese pressure for so many years. Of course, they had little choice but to fight for their existence, as they were "trapped" by their technological basis. Totally dependent on their submarine cables and the necessity of landing sites in countries such as Russia and Japan, the GNTC lacked the possibility of moving in and out of different markets according to the business conditions.

Instead, the GNTC tried to protect itself and its investments through long-term agreements and monopolies, as was the case in both Russia and Japan, and through cartels, joint purses, and strategic agreements, as was the case in the alliance with the Eastern and Associated Telegraph Companies. This strategy was shaped mainly by the LTS character of the company's physical facilities and business organization that restrained its flexibility in dealing with political risk in host countries. To avoid the appearance of "holes" in this system and the possibility of a "domino" effect, the GNTC throughout its whole history did its utmost not to give up cables and stations. In fact, before the outbreak of World War II, the company only gave up two cables: the Japan-Korea cable (in two sequences in 1890 and 1910) and a cable between Norway and Great Britain (1910).[82]

Standing firm on its concessions, the GNTC showed persistence and patience that apparently surprised both the Japanese and the Soviet governments. More generally, the company, since its foundation in 1869, had developed a wide range of managerial capabilities to handle situations like those in Japan and the Soviet Union. These capabilities were more related to international politics and diplomacy than to "ordinary" business, and they were always based on the fundamental principle of "the strictest neutrality," as emphasized by Schønebeck when he appeared before Gosplan in January 1930.

It was a principle that meant neutrality in relation to international politics and the mutual relations between the different states with which the GNTC collaborated. It also demanded neutrality toward the systems and regimes of those countries. So even though the company had no sympathy at all for fascism or Stalinism, it always kept its opinions to itself and avoided any form of public statements. There was simply no alternative, if the company was to survive—and it did.

After World War II the GNTC resumed its business in Great Britain, Japan, and the Soviet Union, while it was expelled from China following the Communist revolution in 1949. In the early 1960s, in cooperation with a Canadian company, the company established a submarine cable link from Great Britain to North America via Iceland and Greenland, and in 1968–1969 it renewed and modernized the trans-Siberian link between Tokyo and London. In 1993–1995 it participated in laying submarine fiber-optic cables in the Baltic Sea connecting Denmark and Russia, and in Far Eastern waters connecting Russia, Japan, and South Korea. The two cable systems were connected via a digital radio chain across European Russia and Siberia.

At the end of the twentieth century, the GNTC finally gave up its international telecommunications activities, and its offices in Tokyo and Moscow were closed down in the years 2000 and 2001 respectively. Today, the GNTC is a multinational company with the name GN Store Nord, manufacturing hearing aids and headsets.

Notes

1. Alfred Schønebeck, Foredrag i Gosplan, January 31, 1930, Great Northern Telegraph-Company Archives (GNTC), Danish National Archive, Copenhagen (hereafter RA).

2. Olivier Coutard, "Introduction: The Evolving Forms of Governance and Large Technical Systems," in *The Governance of Large Technical Systems*, ed. Olivier Coutard (London: Routledge, 1999), 1.

3. George Arthur Codding, Jr., *The International Telecommunication Union: An Experiment in International Cooperation* (Leiden: E. J. Brill, 1952), 13–14.

4. The development that led to the formation of the ITU is outlined in ibid., 15–23.

5. George Arthur Codding, Jr., and Anthony M. Rukowski, *The International Telecommunication Union in a Changing World* (Dedham, Mass.: Artech House, 1982), 3.

6. For the special circumstances of this cable see Poul Thestrup, *Vogn og tog—Prik og streg* (Copenhagen: Generaldirektoratet for Post- og Telegrafvæsenet, 1992).

7. Codding and Rukowski, *The International Telecommunication Union*, 24–26.

8. Jorma Ahvenainen, *The Far Eastern Telegraphs: The History of Telegraphic Communications between the Far East, Europe and America before the First World War* (Helsinki: Suomalinen Tiedeakatemia, 1981), 7.

9. The implications of international telegraphy during the Crimean War are described in Tom Standage, *The Victorian Internet: The Remarkable Story of the Telegraph and the Nineteenth Century's Online Pioneers* (London: Phoenix, 1998), 144–148.

10. Daniel R. Headrick, *The Invisible Weapon: Telecommunications and International Politics, 1851–1945* (New York: Oxford University Press, 1991), 46.

11. P. S. Ring, S. A. Lenway, and M. Govekar, "Management of the Political Imperative in International Business," *Strategic Management Journal* 11, no. 2 (1990): 143. A review of the extensive literature on the concept of political imperative is beyond the scope of this article.

12. Ibid.

13. Ibid., 149.

14. For the founding of the GNTC, see Ole Lange, *Finansmænd, stråmænd og mandariner: C. F. Tietgen, Privatbanken og Store Nordiske, Etablering 1868–76* (Copenhagen: Gyldendal, 1978); Poul Thestrup, *Vogn og tog—Prik og streg*; and Kurt Jacobsen, "Diplomacy and International Business: The Great Northern Telegraph Company 1869–1921," in *Business and Society: Entrepreneurs, Politics and Networks in a Historical Perspective*, ed. Anne-Marie Kuijlaars, Kim Prouddon, and Joop Visser (Rotterdam: CBG, 2000). For a full overview of the GNTC's history including the post–World War II period, see Kurt Jacobsen, "The Great Northern Telegraph Company: A Danish Company in the Service of Globalisation since 1869," in *Between National Histories and Global History*, ed. Stein Tønnesson, Juahni Koponen, Niels Steensgaard, and Thommy Svensson, *Historiallinen Arkisto* 110, no. 4 (Helsinki: Finska Historiska Samfundet, 1997), 179–196.

15. Royal Danish Legation in St. Petersburg to Danish Foreign Ministry, March 21, 1866, Ministry of Foreign Affairs (UM), A3607/1866, RA.

16. Lange and Thestrup are unaware of the financial support from the Danish state. The relationship between Tietgen and the Danish minister of finance is outlined in more detail in Jacobsen, "Diplomacy and International Business."

17. Memorandum on Economic Subsidies to the Danish-Norwegian-British Telegraph-Company, June 4, 1868, fond 1289, opis 1, delo 2827, Russian State Historical Archives, St. Petersburg.

18. Lange, *Finansmænd, stråmænd og mandariner*, 92–93. Thestrup doubts the interference of the Danish king, but the relevant Russian and Danish archives leave no doubt about the Danish royal action. Neither Lange nor Thestrup was granted access to the GNTC archives or Russian archives.

19. Madsen to Tietgen, November 5, 1868, Tietgen Papers, RA.

20. Tietgen's telegraphic vision is analyzed in detail in Lange, *Finansmænd, stråmænd og mandariner*, 70–97.

21 The following is based upon Lange, ibid., 102–145, where he analyzes the negotiations over the Russian concession in full detail.

22. Tietgen, Application for Concession, September 1, 1869, fond 1289, opis 1, delo 2835, Russian State Historical Archives, St. Petersburg.

23. Memo from the Zongli Yamen, June 6, 1870, cited in Erik Baark, *Lightning Wires: The Telegraph and China's Technological Modernization 1860–1890* (Westport, Conn.: Greenwood Press, 1997), 81. The negotiations over the Japanese concession to the GNTC are de-

scribed in Lange, *Finansmænd, stråmænd og mandariner*, 175–177, and Jacobsen "Diplomacy and International Business."

24. The Danish cable laying in the Far East is outlined in Lange, *Finansmænd, stråmænd og mandariner*, 165–216, and in Ahvenainen, *The Far Eastern Telegraphs*, 42–48.

25 The difference in the Chinese reaction to the arrival of the telegraph is a key subject in Baark, *Lightning Wires*.

26. As it appears from the company's official history, published at its twenty-fifth anniversary in 1894, the GNTC itself was very conscious of its exploitation of Denmark's neutral, small status: *Det Store Nordiske Telegraph-Selskab, 1869–1894* (Copenhagen: Det Store Nordiske Telegraf-Selskab, 1894). The GNTC expansion into the Far East, based upon Denmark's geopolitical status, has been outlined in Lange, Ahvenainen, and Baark. Like Lange and Thestrup, Ahvenainen did not have access to the GNTC archives. Baark, however, was given access in the late 1990s. In fact, Denmark never declared itself neutral, but its foreign policy was de facto neutralist.

27. Daniel R. Headrick, *The Tools of Empire: Technology and European Imperialism in the Nineteenth Century* (New York: Oxford University Press, 1981).

28. The early cooperation and controversies between the GNTC and the Pender group is described in Ahvenainen, *The Far Eastern Telegraphs*, 49–52, and in detail in Lange, *Finansmænd, stråmænd og mandariner*, 149–179.

29. The Far Eastern "telegraph war" is the key subject in Lange, *Finansmænd, stråmænd og mandariner*. See also Ahvenainen, *The Far Eastern Telegraphs*, 65–108.

30. Alfred D. Chandler, Jr., *The Visible Hand: The Managerial Revolution in American Business* (Cambridge, Mass.: Belknap Press of Harvard University Press, 1977), 145, 197–200.

31. Headrick, *The Invisible Weapon*, 38.

32. The relations between the GNTC and the Eastern and Associated Telegraph Companies, as well as with the British government, are examined in Kurt Jacobsen, "The Great Northern Telegraph Company and the British Empire 1869–1945," in *Britain and Denmark: Political, Economic and Cultural Relations in the 19th and 20th Centuries*, ed. Jørgen Sevaldsen with Bo Bjørke and Claus Bjørn (Copenhagen: Museum Tusculanum Press, University of Copenhagen, 2003), 199–230.

33. Managing director's monthly report to the board, December 1895, GNTC, RA. Eastern Extension Company, Minute Books of the Board, January 29, 1896, Cable and Wireless Archives, Porthcurno, United Kingdom.

34. Managing director's monthly report to the board, September 1896. GNTC, RA.

35. For details of the Pacific cable project, the formation of the Commercial Pacific Cable Company and the traffic agreements, see Ahvenainen, *The Far Eastern Telegraphs*, 158–174, 178–185.

36. Daniel Headrick and Pascal Griset, "Submarine Telegraph Cables: Business and Politics, 1838–1939," *Business History Review* 75, no. 3 (Autumn 2001): 563. This article contains several misunderstandings about the GNTC.

37. Headrick, *The Invisible Weapon*, 7. It is correct that a new form of "politicizing" seems to have emerged at the turn of the century, but as pointed out in this paper, international telegraphy was "politicized" from its very origin.

38. Cited in Bo Lidegaard, *I Kongens Navn: Henrik Kaufmann i dansk diplomati 1919–1958* (Copenhagen: Gyldendal, 1996), 100.

39. Ibid., for a description of Otto Scavenius's diplomatic career and his entrance into the GNTC board and management.

40. Ring, Lenway, and Govekar, "Management of the Political Imperative in International Business," 145.

41. For the development of the relations between the GNTC and prerevolutionary Russia, see Kurt Jacobsen, "The Great Northern Telegraph-Company and Russia: 130 Years of Cooperation in the Light of the Big Politics," *History of the Motherland*, no. 4 (2000): 44–54; no. 5 (2000): 58–69; and no. 6 (2000): 28–43 (in Russian).

42. The early years of cooperation between the GNTC and Japan have been described in Ahvenainen, *The Far Eastern Telegraphs*; Daqing Yang, "Submarine Cables and the Emerging Japanese Empire," unpublished paper presented at the AIB annual meeting, Charleston, South Carolina, November 20–23, 1999; and Kurt Jacobsen, "In Struggle for Control over the Far Eastern Telegraphs: The Great Northern Telegraph-Company and Japan 1870–1943," in *Transnational Companies 19th–20th Centuries*, Hubert Bonin, Christophe Bouneau, Ludovic Cailluet, Alexandre Ferndandez, and Silvia Marzagalli (Paris: Éditions P.L.A.G.E., 2002), 331–345.

43. The mutual dependency and interests constituted the basis for more than 130 years of cooperation between the GNTC and Russia. See Jacobsen, "The Great Northern Telegraph-Company and Russia."

44; Yang, "Submarine Cables and the Emerging Japanese Empire."

45. Ibid.

46. Ibid.; Jacobsen, "In Struggle for Control over the Far Eastern Telegraphs."

47. Jacobsen, "In Struggle for Control over the Far Eastern Telegraphs"; Yang, "Submarine Cables and the Emerging Japanese Empire." Japan, at this time, had no knowledge of the GNTC's interests in the Commercial Pacific Cable Company that owned and operated the transpacific cable.

48. Jacobsen, "In Struggle for Control over the Far Eastern Telegraphs"; Yang, "Submarine Cables and the Emerging Japanese Empire."

49. Report of the General Meeting held at Copenhagen on April 30, 1910, GNTC, RA.

50. The GNTC's political-diplomatic double game during the Russian Civil War is examined in full detail in Jacobsen, *Den Røde Tråd: Det Store Nordiske Telegraf-Selskabs storpolitiske spil efter den russiske revolution* (Copenhagen: Gyldendal, 1997).

51. Commissariat of Foreign Affairs to Sovnarkhom, December 4, 1920, fond 130, opis 5, delo 74, State Archives of the Russian Federation, Moscow.

52. The Indo-European Telegraph Company also landed cables in the Soviet Union at the Black Sea coast. However, they were used exclusively for a British-operated route between Great Britain, India, and the Far East. In contrast to the GNTC, the Indo-European Telegraph Company was heavily obstructed by the Russian authorities. See Jacobsen, *Den Røde Tråd*, 179–180.

53. See ibid. for the full content of the Soviet concession to the GNTC.

54. Representative in Moscow to managing director, February 12, 1923, UM 64, Dan. 65, RA.

55. The controversies between the GNTC and Soviet Russia in the late 1920s are described in Jacobsen, "The Great Northern Telegraph-Company and Russia," and his "Store Nordiske og Stalin," in Johnny Laursen, Micahel Mogensen, Thorsten Borring Olesen, and Hein Rasmussen, eds., *I tradition og kaos: Festskrift til Henning Poulsen* (Aarhus: Aarhus University Press, 2000).

56. Managing director's Monthly Report to the board, September 1926, GNTC, RA.

57. Ibid., March, May, and June 1933.

58. Ibid., January 1923.

59. Kaoru Hanaoka, *One Hundred Years of Submarine Cables and the Pacific* (Tokyo: 1969), 81 (in English, translated from Japanese by the GNTC).

60 The relations between the GNTC and the Soviet Union during the Stalinist repression are examined in detail in Jacobsen, "The Great Northern Telegraph-Company and Russia" and "Store Nordiske og Stalin."

61. The representative in Moscow to the managing director, July 21, 1937, UM, Dan. 5/53, RA.

62. *Leningradskaja Pravda*, June 11, 1937.

63. Jacobsen, "The Great Northern Telegraph-Company and Russia" and "Store Nordiske og Stalin."

64. Ibid. Commissariat of Foreign Affairs and of Post and Telegraph to Stalin, September 8, 1937, fond 5446, opis 22a, delo 207, State Archives of the Russian Federation, Moscow.

65. The remarkable story of the two Danish telegraph operators and their rescue from the Stalinist terror is still to be written.

66. Appears from several documents in fond 085, opis 22, delo 259, Foreign Policy Archives, Foreign Ministry of the Russian Federation, Moscow.

67. Yang, "From Cooperation to Conflict: Ending Great Northern Telegraph Company in Japan," in Bonin et al., *Transnational Companies 19th–20th Centuries.*

68. Ibid.; Hanaoka, *One Hundred Years of Submarine Cables.*

69. Yang, "From Cooperation to Conflict."

70. Hanaoka, *One Hundred Years of Submarine Cables*, 84.

71. See the collection of translated newspaper articles and the reports from the Danish legislation in Japan in UM 92, Dan.5/5 II, RA. Ironically, the relations in the 1930s between the GNTC and Eastern as well as Great Britain were weakening. See Jacobsen, "The Great Northern Telegraph Company and the British Empire 1869–1945."

72. Yang, "From Cooperation to Conflict."

73. The Japanese actions against the telegraph companies in Shanghai are described in detail in Jacobsen, "In Struggle for Control over the Far Eastern Telegraphs."

74. Danish Ambassador in London to the Danish Ministry of Foreign Affairs, November 12, 1938, UM 92, Dan 5/5N.I, RA.

75. Cited in Jacobsen, "In Struggle for Control over the Far Eastern Telegraphs."

76. Ibid; Yang, "From Cooperation to Conflict."

77. Managing director's monthly report to the board, March 1940, GNTC, RA.

78. In countries such as Denmark, Great Britain, and Germany the initiative to erect domestic telegraph networks had been taken by private companies, but their systems were soon nationalized.

79. Headrick, *The Invisible Weapon.*

80. For instance, all telegraphic communications between the Comintern and the Danish Communist Party as well as with other Western European Communist Parties were handled "via Northern" until the mid-1930s.

81. Yang, "From Cooperation to Conflict."

82. For the Japanese takeover of the Korean cable, see Yang, "Submarine Cables and the Emerging Japanese Empire"; and Jacobsen, "In Struggle for Control over the Far Eastern Telegraphs." For the takeover of the GNTC's cable Norwegian-British cable by the Norwegian state, see Jacobsen, *Den Røde Tråd.*

8

Je t'aime, moi non plus: The Development of Atlantic Submarine Cables and the Complexity of the French-American Dialogue, 1870–1960

Pascal Griset

La vapeur, le télégraphe sont devenus les instruments du règne.

Discours et plaidoyers choisis de Léon Gambetta (1868)

Since the 1830s telegraphy has been a state-owned monopoly in France.[1] Nevertheless, for specific technologies or markets, private companies have been allowed to operate new services.[2] While this possibility emerged only exceptionally for domestic networks of telephony (1878–1888) and radio broadcasting (1923–1940), in international telecommunications nongovernmental services were more common. Between 1869 and 1954 several private companies were allowed to operate cables and radio,[3] though under strict state control.[4] Likewise, some foreign companies were allowed to operate communication lines entering or leaving the country. However, neither French nor foreign companies were allowed to operate their own lines on French territory; they had to connect their networks to the network of the ministry of Postes et Télégraphes (P&T), later Postes, Télégraphes et Téléphones (PTT). This mixed system of state-owned monopoly and private interests gave rise to a complex system including three main actors: American and British companies, French private companies, and French state administration.[5]

The specific context of international telecommunications forced institutions that usually worked behind closed doors under radically different principles to cooperate with one another.[6] We are thus in a position to observe behavior that would reappear during the process of deregulation in the 1980s. Because of their international role and their privileged geostrategic position[7] on the Atlantic (the "Voie Royale" of international telecommunications and transportation), France and Great Britain were the essential interlocutors for the United States in its attempt to establish fruitful relations with Europe.[8] The French-American dialogue was characterized by periods of strong tensions but also by mutual conviction that cooperation could on no account be jeopardized. This dialogue evolved with the ascent to power of U.S. companies and the growing international ambitions of the United

States. After a period of mutual discovery, conflicting points of view and interests began to assert themselves.

French Interests: A Pawn in Anglo-American Rivalries

> The day was one of the most beautiful of summer, calm and bright. . . . everything in nature seemed to look propitiously upon the great work about to be completed. . . . When the heavy work had been finished, and the assembly realized that at last the Empire of France and the Republic of America were united by a living, pulsating artery, another involuntary shout of gladness burst from every mouth.[9]

On July 23, 1869, installation of the first telegraphic cable directly connecting France and the United States was completed—without any license obtained from the American government.[10] On July 10, 1869, while the steps toward installation were already well under way, the secretary of state wrote to the count of Taverney to announce his surprise at such carelessness: "As the projectors of the enterprise now in progress have not received or awaited the permission of this government, either to the establishment of this telegraphic connection or to its use for such intercourse with the coast of France. . . . In thus distinctly notifying you that in the absence of the assent to or concurrence in the proposed telegraphic connection of the Government of the United States, the proceedings of the parties interested must be taken in submission to the authority of this Government in the premises."[11]

In fact no legislation on these questions existed, with the result that the administration was not able to take any measure to prevent the landing of the French cable. Regularizing forces would soon intervene to dissolve this impediment. President Grant himself took a hand in the case. Adopting a precedent-setting attitude, the president refused to grant a license unless the company gave up a privilege of monopoly that seemed to have been granted by the French administration. He added a condition that, in the future, American companies would be granted the same licenses in France. On March 31, 1870, the French ambassador in Washington received the agreement of the government on the question. He even proposed that the two governments jointly fix a maximum rate for companies operating this line. Even if this last idea, although approved in principle by Grant, was not accepted, the company's position was finally clarified.[12]

Commenting some years later on his decision, Grant explained: "I was reluctant to deny to the great interests of the world and of civilization the facilities of such communication as were proposed. I therefore withheld any resistance to the landing of the cable on condition that the offensive monopoly feature of the concession be abandoned."[13] Thus from this case a doctrine emerged: reciprocity of landing rights; prohibition against a company merging with a rival company in order to maintain high prices; priority for official messages; and the right of states concerned to fix a maximum rate for telegrams. Such were, for the next decades, for lack of real legislation, the necessary requirements for obtaining a license to land in the United States.[14]

The first encounter between France and America in telecommunications had no real consequences. The French project was a production of the banker Emile d'Erlanger;[15] engaged in the most acrobatic of financial arrangements and in often suspect businesses, Erlanger's company was in fact only the representative of British interests desirous of landing a cable on French soil.[16] This first company fell quickly under total British control. The situation remained unchanged for ten years,[17] until in 1879 Senator Pouyer-Quertier created a new company, the Compagnie Française des Câbles Télégraphiques (CFCT), to at last give France the independent transatlantic communications that the country needed.[18] Soon this company in its turn was forced to give up any autonomy. After a terrible tariff war the young company, completely exhausted, had to integrate with the pool that associated American and British interests under English leadership.[19]

In the mid-1880s John William Mackay's and James G. Bennett Jr.'s newly created company, Commercial Cable, radically changed the balance of forces. The American Mackay had the financial wherewithal to destroy the British-led monopoly and survive the tariff war that the cartel instantly opened. In order to force his opponents to negotiate, Mackay understood very early that he should single out the French company as the weakest partner in the pool.[20]

He put, by more or less secret methods, strong pressure on the CFCT, with the aim of seeing the French company abandon the British alliance and join the young American challenger. Such a movement would offer him an ally and would carry a cruel blow to the pool. His agents were soon able to control enough shares for him to acquire a very significant influence in the General Assembly of the share holders of the French company. In order to enhance his position, Mackay's agents also used more disreputable methods. The administrators of the French company discovered, for example, during the summer of 1885 that Mr. Renard, second-in-command of the office of taxes, delivered in exchange for money important documents, in particular customers' lists, to Count Arthur Dillon, a consultant working for Commercial Cable.[21]

This work of undermining produced its first fruits during the General Assembly of 1886. The debate between the supporters of continuity—that is, staying with the British—and of change—joining a new alliance led by Commercial Cable—was especially contentious. Management, which favored the status quo, was intensely questioned by a group of shareholders. As their action was noisily supported by agents that Mackay paid to destabilize the assembly, they were able to obtain a real change in the composition of the board of administrators, giving many more positions to persons willing to alter the company strategy.

The new administrators began by undertaking a detailed audit of past reports. They denounced, for example, the fact that official documents were entirely missing for 1881.[22] Beyond that, upsetting Pouyer-Quertier's "British strategy" was their real goal. The new board had no difficulty proving that the tariff war, originally supposed to be "short and cheap," was extremely costly to the French company. The situation for the British camp became even worse when it was proved that the accounts were in fact falsified in order to hide the terrible consequences of the English alliance. The count of Lambertye, a close friend of Pouyer-Quertier, was forced to resign. The former minister's position was thus appreciably weakened.

His enemies organized, at the last minute, a so-called extraordinary board of directors on December 31, 1886. Informed too late to counter the action, Pouyer-Quertier was not able to attend. At the conclusion of a meeting that did not last one hour, a decision was taken just before midnight to denounce the agreements binding the company to British interests. The president resigned shortly thereafter.[23]

The second part of the plan was revealed at the beginning of February, when the company announced the signing of a treaty with the Commercial Cable Company. It was an advantageous agreement that assured the French company the free use of a recent cable as a kind of reward.[24]

The negative consequences of the breech of the British agreement appeared, however, also very quickly. First, on the legal side, a summons was delivered to the company, in the name of the Anglo-American Telegraph Company, to appear in front of the commercial court of the Seine, where they were ordered to resume the execution of the contracts under penalty of F 5,000 for each day of delay. There were consequences also for exploitation under the new arrangement, because Western Union refused to return its landline to the French company. A new line would have to be built from Cape Cod to Taunton. An independent desk would also have to be opened in New York.[25] Compounding problems, the Postal Telegraph Cable Company refused to let Commercial carry traffic for the CFCT when its cable was cut.

Far from dismissing the affair as an accounting matter of profits and losses, the pool counterattacked energetically. An impressive coalition—Anglo-American, Western Union, and even the British General Post Office—was attacking the small CFCT! Far from realizing a happy future under Mackay's umbrella, the French company was in fact confronted with a real nightmare. Relations between the various companies gradually normalized, according to a rhythm determined by the Commercial Cable Company: after its negotiations with Anglo-American, the CFCT played only an incidental role. Gradually management of the French company passed under American influence. With the appointment of Count Dillon,[26] a Commercial man, as general manager, there was no longer any ambiguity regarding the independence of the French company.[27]

An agreement reached in July 1888 established the new balance of power. Replacement of the English-led alliance by an American-led alliance resolved none of the French company's problems, though. Nevertheless, and paradoxically, Mackay's efforts had brought to fruition some of the seminal ideas that had motivated creation of the CFCT in 1879. Certainly the road had been tortuous, but Britain's Atlantic monopoly had been seriously assaulted, and the British were forced to negotiate and to accept important rate reductions. This reconfiguration had not, however, redounded to the benefit of France.[28] The Atlantic cartel was from now on a structure with two heads.[29] Mackay had destroyed the former agreement. The previous cartel, based domestically in the United States on the landlines of the Postal Telegraph Cable Company,[30] was certainly less powerful than the Western Union–Anglo-American pool. And the new cartel was no less capable of treating Mackay as an equal. The French interests, on the contrary, were totally marginalized in the new system.[31]

1914–1918: Allies in the Conflicting Cultures

A new period of limited relationships opened between the French and the Americans during the First World War.[32] Transatlantic telecommunications took on vital importance during this period, as was widely anticipated.[33] Even though they were allied in fighting Germany, both nations found motives for dispute. Censorship, which had also been anticipated, notably by the British, was in effect from the opening of the conflict, the equivalent measures being taken in France.[34] American complaints concerning French censorship generated voluminous correspondence between the two countries' governments. Censorship indeed seemed to hamper considerably the intercourse of American companies with Europe, particularly Switzerland and the Iberian peninsula. The French administration's explanations to the Americans reflected a desire to intercept any message to or from a company suspected of business connections with Germany. A superficial examination of index cards communicated to the U.S. embassy is enough to indicate that the interceptions were a consequence of intensive work, cross-checking a variety of different sources, with special emphasis on radio transmissions. "Constant relations with Germany," "suspect relations with Germany"—vague allusions to unknown companies constituted the most frequent motives.[35] The case of Hoffmann-Laroche, for example, turned out to be particularly sensitive. Numerous messages related to the chemical manufacturer were intercepted by France, for "enigmatic" reasons.[36] The very stiff rules originally established by the French were gradually somewhat relaxed to take into account criticisms. In October 1915 the minister of war accepted a proposition from the foreign secretary transmitted by the U.S. ambassador in Paris. According to this proposition, a sender had to be warned when a determination was made to block delivery of a telegram.[37]

After November 11, 1918, the president of the United States asked the secretary of state to intervene with the French and British authorities to stop any censorship of telegrams of the American press agencies. The argument for military security could no longer be supported, and there was a risk of alienating public opinion in the United States if the practice continued.[38] Even if this point was accepted, censorship of private mail continued after the end of the conflict, provoking very strong American protests. These came to an end only with the abolition of the system, quite late, on August 15, 1919. This excessively long delay left smoldering for many months suspicion over the practice of the French and British administrations. It even seems that censorship was restored for a little while in October 1919 (claims of Standard Oil and General Motors on October 15, 1919).[39] As late as March 1921, a company in Pennsylvania complained about the nonliberation of a telegram to one of its Parisian correspondents because of French objections: "Will you kindly inform us why there should be any reason for non-delivery of a cable of a commercial nature to a French citizen, even though of Russian birth, without any explanation to us? To our knowledge Monsieur Foulon is a French citizen and has no connection or affiliation of any kind with the Soviet Government in Moscow."[40]

The main effect of wartime censorship was that for customers it cast doubt on the cable companies' honesty and respect for privacy. Was not military censorship above all political, even economic? After the conflict, the president of Congress's Board of Inquiry on submarine cables wondered about the excesses of censorship during the war: "Do you know that during the war commercial messages between American merchants or dealers, and merchants and dealers in neutral countries were taken off the wires and entirely censored by the British and French governments? . . . Do you know, as a matter of fact, whether those messages between merchants who do business in American countries and traders in neutral countries were turned over to British traders and they were permitted to do the business?"[41]

More than France, however, Great Britain was suspected, though no proof of this has been found. Relations between American cable companies and the French PTT also suffered from the effects of the war. Of particular concern was official traffic, which made up a large amount of the activity of the U.S. companies.

The concession granted to Commercial on December 17, 1884, anticipated the company's free transmission of official French governmental telegrams to the United States. The French government had used this privilege in moderation, but the war aggravated the problem, with this type of traffic multiplied by ten between 1913 and 1917. The service became a very heavy load for the American company, which felt it all the more inequitable because, according to its information, the French company paid a terminal tax equal to 25 percent of the normal rate for French government traffic[42] while Western Union in the same scenario paid 50 percent of the normal rate.[43] Clarence Mackay noted that France was the only government to impose such a benefit to its company, with the governments of the United States, Portugal, Japan, and China paying 50 percent of the normal tax.

The British government, which had stipulated in the agreement granted to Commercial that it should receive conditions equivalent to the most advantageous granted to any other foreign government, judged the conditions imposed by France on Commercial "inequitable." "Steamships are paid for carrying mails. The French Government certainly has not required steamships of the British or American lines to carry freight or mails of the French Government free. . . . There is no reason why the Commercial Cable Company should be discriminated against by requiring it to transmit Government messages gratuitously."[44]

Commercial's complaints, however, went beyond the problem of governmental telegrams. The company wanted the right to represent itself directly to the French public. It supported its demand on the basis of French-American agreements that, on the occasion of the first connection between the two countries, bound the landing permit of the French cable on American soil to complete reciprocity of advantages given to the French company in the United States. The French company had seven desks in the city of New York, so Commercial demanded seven desks for itself in France.

This reciprocity should also mean having the freedom to open lines on French soil to connect these desks with the transatlantic cable, as the French company had done in the United States. Clarence Mackay added, "Under all these circumstances we feel justified in insisting upon a modification of our permit from the French Government, so that our Company shall be treated the same as Anglo American Telegraph Company, a British cable company (now leased to the Western Union Company) and so that the Commercial Cable Company shall not be singled out for inputs and illegal discrimination in violation of the international agreement between the French and American Governments."[45]

In the eyes of the PTT administration, this demand carried the germ of questioning its monopoly. In January 1918 a compromise proposed by the PTT was rejected by Mackay,[46] and the American entrepreneur reacted with more and more excess. The messages of the French consul, for example, were characterized as "contraband."[47] The ultimate stage in the escalation was reached when, on August 11, Commercial refused French official telegrams. By interrupting the connections even as battles raged on French ground, Mackay made an impolitic error. Confronted with this escalation, the State Department, which had at first supported Mackay,[48] was convinced that an agreement should put an end as quickly as possible to this painful affair. On August 30, Robert Lansing wrote to Mackay to make him understand that the agreement proposed by the French (payment of 50 percent of the taxes over 60,000 words and global renegotiation six months after the end of the conflict) was acceptable. "It is suggested, therefore, that you may desire to instruct your representatives in Paris that entire question is now in hands of Department in order that your company and Department may not be working at cross purposes."[49] Isolated by American diplomacy, Mackay was forced to give up. An agreement was signed, and the official telegrams were again accepted free of charge.

The dispute had been settled only in a very temporary way, in the context of the emergency created by the German offensive. The problems reemerged after the armistice. But, after new exchanges of fire between Mackay and the administration, an agreement on July 30, 1919, at last allowed the settling of accounts. The Foreign Ministry could finally inform the embassy of the United States that the affair was closed.[50]

Other problems arose, though. The monetary exchange rate constituted one of the most painful cases in the immediate postwar years. The franc had been greatly overvalued during the conflict, and afterward it depreciated even more. Paid in francs, without taking into account the French currency depreciation, Western Union and Commercial were hurt.[51] The

French administration apparently tried to buy time in this affair. It even cited the nonmembership of the United States in the International Telegraph Convention to justify its attitude, although it behaved in the same way toward the British company Eastern Telegraph, based in a country that *was* a signer of the agreement.[52] The affair took an unexpected turn when English banks opened temporary agencies in Paris so that their telegrams to the United States could go by way of France and they could benefit from the very advantageous exchange rate. They even went so far as prepaying the replies. This problem was gradually solved by the introduction of a coefficient of rate increases that took into account the depreciation of the franc. The bad faith with which the French administration had handled the affair did not, however, favor the institution of a reliable climate with the Americans.[53]

New Framework and New U.S. Rhetoric for a Wider Strategy

After the First World War, plans for development of the American cable companies came into conflict with European government administrations. Control of the transatlantic connections was logically followed by conquest of the European ground networks. The weight of the American continental network had inevitably led to the fall of the Atlantic Telegraph Company system to the west. Between San Francisco and the European coast a kind of integrated network under American control developed. In continuing its eastward expansion, this new system began again at the shores of continental Europe. In order to continue to spread east, the American companies had to develop noncommercial arguments. The links between the United States and Europe were operated directly or indirectly by American companies; nevertheless, all these connections depended physically on landings situated in Great Britain or in France. Diversification of the territories where American cables could connect to continental Europe would constitute the first step in a new phase of development. The problem had been momentarily addressed by the connection of two cables between Germany and the United States via the Azores, without relays in France or in England. Allocation of these cables to those two countries again placed the U.S.-Europe telecommunications under direct U.S. control.

At the same time, development of radiotelegraphy and the ambitious strategy of Marconi caused U.S. interests to be concerned about the return of British hegemony in the field of intercontinental telecommunications.[54]

This problem was underscored by Walter Rogers, a negotiator during the major international conferences, who called for direct connection between the United States and most of the European countries. "My own feeling is that the Atlantic is largely taking care of itself," he said, "except that we ought to have a direct cable, if possible, with the Scandinavian countries and have a cable, through relay points, if possible, to the Mediterranean bases . . . one or more to Germany and one or more to Italy."[55]

This project found an echo in plans developed by Western Union. In 1921 the company determined that it would establish direct connections with Holland, Norway, Sweden, and Germany. In that way the company would make its contribution to the international policy of its country. By freeing itself from French or English landlines, Western Union would also see its share of the terminal tax appreciably increased. Thus the political will of the U.S. government and the commercial ambition of the cable companies joined to make Europe a priority objective.

After a few months, however, the already sensitive competition with radio[56] and a less favorable economic context imposed much less ambitious objectives. It was no longer possible to believe in a business plan based on new and expensive cables landing in Germany or Italy. After euphoria had subsided, American companies were forced to recognize that Brittany and Ireland were still the most efficient landing points. The situation was even "simpler" than it had been. Because of the political situation in Ireland, a landing there was too risky and much less attractive than in the past.[57] They had to admit that France—and its administration—could not be bypassed!

It was advisable, however, to make France understand that it had to join an organization of telecommunications with a model no longer based on the legitimacy of governmental administrations but managed by the power of companies. PTT doctrine had not evolved in this direction. It still consisted of granting to foreign companies conditions of exploitation identical to those granted to French companies.[58] American companies, for their part, wished to be treated just as the U.S. administration treated the French company. The dialogue was made even more difficult by the key U.S. demand of opening desks directly managed by an American staff.

The American companies had signed contracts that were surprisingly agreeable to the French administration.[59] Acceptable at the time of signature, these old conventions became an annoyance as the ambitions of the companies increased. Any modification implied a renegotiation and had to base itself on arguments that went beyond particular interests. The common good—an implicit theory behind any progress in negotiations related to international telecommunications (progress being something that an archaic French administration would automatically oppose)—had to be integrated gradually into the language. Two main polemical axes gave support to this approach. On the one hand, American professionals saw the French administration as incompetent. On the other hand, and in contrast to the British commitment to competition, the French administration seemed incapable of admitting to the principles of a free market economy.

Americans' assertion of their superiority in matters of organization, working methods, and technique became more and more evident as the debate continued. The arguments moved increasingly from the commercial domain to the "nobler" field of technical quality. "We regard operation by our own operators and continuous handling as absolutely *necessary* to good service," said Newcomb Carlton in September 1920. "A cable may be, often is, sent by our operators from the Rue de Grenelle in Paris to our office in London in a couple of minutes . . . but if it is held up, as it frequently is, in the C.T.O for hours before being handed by the French functionaries to our operators, all our efforts go for nothing."[60]

Clarence Mackay developed the same arguments against the attitude of the French administration: "We can not regard with satisfaction, conditions which confine us to the mere physical landing of a cable and compel us to rely upon the government telegraph systems to perform the inland operations, and to deal with the public in the collection and delivery of messages."[61] The Associated Press was very critical of the lack of speed of the PTT administration and advocated for American cable companies' opening desks on French territory and assuming responsibility for the whole activity: "We find American operation much more efficient; at least we prefer it."[62]

Thanks to statements like these, the American companies were no longer presented as only defending their interests and trying to increase profits; now they were representatives of a universal social institution, basing their superiority on objective criteria of performance in their service to the public, without consideration for international borders and the sovereignty of nations.[63]

To increase the import of this argument, it was necessary to make the French administration "feel guilty," while demonstrating that, with a little willingness, its situation was not desperate. In order to educate the bad French pupil, it was necessary to enlighten him to the right attitude, of course meaning the British example. Carlton explained, for instance, that around the world—"except in England"—the telegraphic network was lower in quality than in the United States.[64]

"Before the war," added Mackay, "our company had made all the preparations necessary for the installation of another transatlantic cable from New York to Europe, via the Azores. Because of the little liberal conditions imposed by France, it is very possible that we decide to choose finally Great Britain."[65] France was presented, in the rhetoric of the American companies, as the dunce of the world telegraph system. It is true that these opinions were not groundless. Objectively, American telecommunications systems were globally very much ahead of those proposed in the French PTT. The comparison with Great Britain, on the other hand, seemed more specious.[66] If there were differences between British and French domestic networks, they were relatively small.

Curiously, in documents describing conflicts between the American companies and the General Post Office, points of view on Great Britain were very different. For example, in January 1920 Mackay summoned all his eloquence to denounce the blackest of the Post Office's intentions: "It is not beyond the realm of possibility that ruinous rates may be forced upon us for the specific purpose of compelling us to sell our lines between Great Britain and Canada to British interests, supported by the British Government, to destroy the present control by American companies of the Atlantic cables so that they may again pass back into the hands of the British, because ever since the Western Union Company took over the control of the two British companies it has aroused a considerable amount of feeling on the part of the British authorities and public men, and there has been a good deal of discussion from time to time as to Great Britain's situation in regard to Atlantic cable communication."[67] The "British reference" was useful only to criticize the French attitude and then conveniently disappeared.

Support from the Department of State

Despite the energy of Mackay's attack, the French attitude remained the same. In consequence, Mackay enlisted the support of the U.S. government starting in November 1919.[68]

By asking for the landing of a new cable in Le Havre, Commercial was the first to give up direct connections to Central Europe via Germany, Holland, or Italy. The request, presented on November 26 to the French administration, received a relatively fast answer. On February 9, 1920, Mackay was informed that the right to land a new cable on French coasts was granted to his company.

This was not enough to satisfy Mackay, who demanded, in addition to landing rights, a license to open specific desks in France and a renunciation by the French government of their right to free transmission of official telegrams. The French administration did not accede to these two demands and only granted landing rights. In his usual prolix fashion,[69] Mackay then sent an eighteen-page report to the Department of State to demonstrate the injustice of the situation.

The general spirit of the text Mackay submitted was, however, appreciably different from those written during the disputes of the years 1917–1918. During his earlier exchange of fire with the French administration, Mackay's position had been doubly weakened. He had threatened an ally of the United States during the war, and he supported his claim by citing another American company, Western Union, as one of his opponents. By 1920 everything was changed. In place of the solidarity among allies, privileged during the conflict, a more nationalistic mood prevailed. The image of France had also evolved. For many Americans, the brave ally had become a new continental power—unfair with Germany, asserting hegemonic claims, and, worst of all, failing to pay its war debts.

In this context, Mackay was able to appear as the defender of U.S. interests. "The question involved," he explained, "embraces not merely cable traffic between the United States and France, but also cable traffic between the United States and Italy, Spain, Austria and Central Europe generally."[70] As a bridge to the whole of Europe, France should not be allowed by the U.S. government to exploit its monopoly position by limiting the freedoms of American companies abroad. It was the whole of American business that Mackay claimed to defend, not just the interests of Commercial: "Who controls cables holds an essential advantage to control the business."[71]

Beyond these remarks, the requirements of the Commercial Cable Company were very exactly formulated:

- the right to desks open to the public without intervention by agents of the state,
- the ability to connect these desks freely to the whole of the company's worldwide network,
- the right to keep these desks at the most convenient hours, including, for some of them, day and night.

Under these conditions, the Commercial Cable Company should not pay any more terminal taxes to the French administration; it should pay only for the rental of the wires being used.[72]

These demands were picked up by Western Union. Although its objectives were similar to those of Commercial, Western Union adopted a more discreet attitude, wanting to show itself supportive of the French administration.[73] "We have no wish to engage in a local telegraph service of even the most restricted character in France or to compete with the French Telegraph Administration in any respect. . . . There is not involved any element of competition with the government monopoly or of economy effected at the expense of the Government Administration, but on the contrary the revenues of the French Telegraph Administration should be increased and the public will be much better served." The unique objective of Western Union was, according to its president, "to improve the character of the cable service between this country and France."[74]

The 1920 Washington Preliminary World Conference on Electrical Communications boosted the campaign led by the two companies. A campaign of extremely pugnacious lobbying, based essentially on mobilization of U.S. chambers of commerce on behalf of the telegraph companies, was directed this time at the Department of State. Numerous chambers of commerce wrote to the Department of State to support Commercial and Western Union. Typical was the message from Philadelphia:

> The only two European countries with which the United States is in direct cable communication are England and France. In England, American cable companies are permitted to operate their own offices with their own trained employees who are specialists in the handling of difficult code correspondence, and they receive messages directly from and deliver them directly to the public, thus insuring a quick and specialized service which is readily adaptable to the needs of particular lines of business, and not subject to mistakes and delays through the interposition of other agencies. In France the cable companies are compelled to transact their business through the medium of the French Telegraph Administration. . . . No reference to the inferior quality of European Government operated service, as compared with the service to which American business men are accustomed, is necessary at this point. The inadequacy and unsatisfactory character of the cable communication with France, with the French Post Office as an intermediary, is painfully evident to everyone who has recently tried to do business with France.[75]

Similar messages reached the Department of State at virtually the same time; from Denver to Sioux Falls, dozens of complaints converged in surprising simultaneity. The Rotary Club of New York also saw fit to support the American companies. Nevertheless, the campaign was ineffective.

Congress's passage of the Kellogg Act in May 1921 radically changed the nature of the problem. By at last officially declaring the principles behind granting licenses for foreign cables to land in the United States, this legislation would become an effective means of applying pressure against France. Through misunderstandings and delays, the situation deteriorated in the early months of the following year. The opening of radio links between France and the United States, operated by a French private company (Radio France) through an arrangement with RCA, further aggravated the American telegraph companies.[76]

These pressures—the passage of the Kellogg Act and the opening of wireless connections—finally forced the French administration to revise its position. It decided at the end of the summer of 1922 to grant Commercial the same opportunities as those granted to Radio France. In spite of a desire by the American companies to reject these partial propositions and to continue an attitude of confrontation, the Department of State seized the occasion to put an end to this long conflict.[77]

Within the administration of the PTT, spirits improved appreciably. In a note of April 1923, the vice president of Western Union noticed, "Except for the one or two men in the Telegraph Administration, the idea of giving the Companies facilities in respect of their own offices has now grown to be regarded as quite reasonable. . . . It is curious to find that there has been amongst the rank and file of the staff such a complete turnover of opinion in something under two years."[78] Both American companies were authorized to open Parisian desks in June 1923. All these services were opened to the public on November 28, 1923.

At the end of 1923 relations between the American companies and the French administration had relaxed appreciably. The position of Commercial Cable was very sharply improved thanks to the laying of a new transatlantic cable landing at Le Havre. The new state of affairs would persist for another thirty years.

Epilogue

Though undermined by the consequences of World War II, including decolonization and political instability, France managed to maintain its presence in the field of transatlantic telecommunications. In 1954 the government created a unique state-owned company called TéléFrance, which combined the French intercontinental cable and wireless networks. The reaction of the American telecommunication companies demonstrated that, for them, French interests always presented an embarrassment. The new entity indeed seemed to assert itself more as a competitor than as a partner.

The personality of the new boss of TéléFrance, Maurice Lhermite, seems not to have worried American professional circles, which feared a more aggressive French competition. Indeed, the first reports on Lhermite reaching the U.S. embassy presented him in a curious way: "He has the hard head, he can be violent, he has at least an unfriendly attitude toward the United States." This view of the man was undone by direct meetings at the U.S. embassy in Paris.[79] Reforms organized by the French administration had, however, the consequence of chasing away American companies from

French soil. There was nothing more in France than a state-owned monopoly and so no room for a private company—French or American. After September 1, 1958, the desks of the American companies were forced to stop any activity on French territory.

These successive stages led to the creation of France Câbles et Radio in 1959. All the radio links and cables under French control were integrated into a single structure. From this date privileged relations were established gradually between the PTT and AT&T. In particular, they allowed the French administration to be associated with the development of the transatlantic telephone (TAT) cables and with the opening of satellite connections.

Conclusion

Within these confrontations of French and American interests, several themes can be perceived. First, there was increasing interpenetration of political and economic questions, with companies leaning on their countries' diplomatic bodies to further their own interests. Second, aside from any legal arguments, there was an assertion of intercontinental telecommunications as a natural area of expansion for the new American power. How is one not to feel, in reading the pleas of Commercial and Western, an assertion of American superiority and a certain annoyance with the French attitude? Third, an impression of mutual incomprehension is indicated by the tone of the various notes and letters. When the veneer of official terminology is stripped away—and in reports of informal meetings—one notices that the antagonisms are quite strong. Reporting to Secretary of State Hughes in November 1922, Newcomb Carlton cannot hide his exasperation over a meeting between a representative of Western Union and a French civil servant: to Western Union's questions on the status of Radio France, the civil servant answered, "Nous pouvons faire ce que nous plait chez nous!" Later, when the representative of Western Union informed him that, in the event that the French attitude remained unchanged, the landing licenses for French cables and the traffic contract with RCA could be canceled, the man of the PTT retorted, "Les Américains ne sont pas assez malins pour faire cela. Il n'y a rien à craindre."[80]

Finally, antagonism in spirit was just as important as in material interests because, concretely, the financial stakes were not very important. Beyond the economic issues, there was a lot of sovereign power at stake. The American logic of an open space for communications, where the best company offers the best service to the public, opposed that of France, which was based on a much more political, even kingly, vision of communications. France stuck adamantly to its positions, giving up to the American companies only on points already granted to Radio France. The government remained faithful to its doctrine and continued to control its landlines, which were doubtless the ultimate objective of the American companies. The French will to preserve this autonomy could seem derisory in front of the power of the American companies, which had been internationalized during the interwar period. Nevertheless, it established the basis for a recovery that allowed French industry to develop strongly in the 1970s (Alcatel)[81] and the national operator (renamed France Télécom) to survive and later, in the years 1980–1990, to develop into a high-quality network.[82]

Aside from telecommunication problems, the dialogue turned out especially difficult between the two sides when they based their reasoning on principles—not to say ideologies—that were quite different. Another indication of a tumultuous but strong secular friendship was underlined by André Malraux during a dialogue with Charles De Gaulle:

> *André Malraux*: Je savais bien ce que pensait le président [John Kennedy]: les États-Unis ne peuvent fonder leur politique européenne sur la France, et ils ne peuvent pas tenir la France pour négligeable, parce que les Français sont toujours capables d'inventer on ne sait quoi: ils ont bien inventé le général de Gaulle! . . .
> *Charles de Gaulle:* Si les États-Unis deviennent consciemment maîtres du monde, vous verrez jusqu'où ira leur impérialisme. . . .
> *A. Malraux*: Kennedy a certainement voulu une action historique, pour lui et pour les États-Unis, mon général. Mais ce n'était pas rien que de concevoir l'action du plus puissant pays du monde sans la concevoir comme un impérialisme.
> *C. de Gaulle*: Je vous ai dit ce que j'en pensais. . . . Vous avez eu l'occasion d'assister à leurs grandes réunions de hippies?[83]
> *AM*: I knew very well what the president was thinking: the United States cannot base its European policy on France; yet it cannot ignore France, because the French are always capable of inventing whatever is needed: after all, they invented general de Gaulle! . . .
> *CdG*: If the United States comes to the point of feeling itself the master of the world, you will see how far their imperialism will go. . . .
> *AM*: Kennedy certainly wanted a historic action, for him and for the United States, my general. But it is no small thing to consider an action taken by the most powerful country in the world, and not to consider it as an act of imperialism.
> *CdG*: I told you what I thought of it. . . . Have you had a chance to go to one of their great hippie gatherings?

Notes

1. E. Delfieu, "Le monopole télégraphique et téléphonique" (doctoral dissertation, Université de Montpellier, Nîmes, 1918).

2. Far from being an instance where, as in Germany, the monopoly was almost uncontested, in France there were some acrimonious debates on the legitimacy of the PTT's monopoly. See for example Henri Fayol's *L'incapacité industrielle de l'État: Les P.T.T.* (Paris: Dunod, 1921).

3. L.-J. Libois, *Genèse et croissance des télécommunications* (Paris: Masson, 1983).

4. Pascal Griset, "Communication et non-communication: Les paradoxes d'une histoire des monopoles," *Crises*, May 1995.

5. Daniel R. Headrick and Pascal Griset, "Submarine Telegraph Cables: Business and Politics, 1839–1939," *Business History Review* 75 (Autumn 2001): 543–578.

6. Pascal Griset, *Technologie, entreprise et souveraineté: Les télécommunications transatlantiques de la France (1869–1954)* (Paris: Institut d'Histoire de l'Industrie, Editions Rive-Droite, 1996).

7. Pascal Griset, "Technical Systems and Strategy: Intercontinental Telecommunications in the First Quarter of the Twentieth Century," in *The Governance of Large Technical Systems*, ed. Olivier Coutard (London: Routledge, 1999), 58–72.

8. Pascal Griset, "Un fil de cuivre entre deux mondes: Les premières liaisons télégraphiques transatlantiques," *Quaderni* 27 (Autumn 1995): 97–114.

9. *The Landing of the French Cable at Duxbury, Mass., July, 1869* (Boston: A. Mudge & Son, 1869), 11–12.

10. The concession granted by R. Feuton, governor of the state of New York, for 300,000 francs, had no legal value.

11. H. Fish to Comte de Taverney, July 10, 1869, in Senate Committee on Interstate Commerce, Cable-Landing Licenses, *A Bill to Prevent the Unauthorized Landing of Submarine Cables in the United States: Hearings on S. 4301*, 66th Cong., 3rd sess., December 15, 1920, to January 11, 1921.

12. Senate Committee on Interstate Commerce, *A Bill to Prevent the Unauthorized Landing of Submarine Cables*, 7.

13. Lester G. Lindley, "The Constitution Faces Technology: The Relationship of the National Government to the Telegraph, 1866–1884" (PhD diss., Rice University, 1971).

14. This incident gave birth to what came to be known as Grant's Doctrine, the essential elements of which are quoted in Senate Committee on Interstate Commerce, *A Bill to Prevent the Unauthorized Landing of Submarine Cables*, 7:

I. No line should be allowed to land on the shores of the United States under the concession from any power which does not admit the right of any other line or lines, formed in the United States, to land and freely connect with and operate through its land lines.

II. No line should be allowed to land on the shores of the United States which is not by treaty stipulation with the Government from whose shores it proceeds or by prohibition in its charter or otherwise to the satisfaction of this Government, prohibited from consolidating or amalgamating with any other cable-telegraph line, or combining therewith for the purpose of regulating and maintaining the cost of telegraph.
III. All lines should be bound to give precedence in the transmission of the official messages of the governments of the two countries between which it may be laid.
IV. A power should be reserved to the two governments, either conjointly or to each, as regards the messages dispatched from its shores to fix a limit to the charges to be demanded for the transmission of messages.

15. See Alain Beltran and Pascal Griset, *La croissance économique de la France au XIXème siècle* (Paris: A. Colin, 1988); Patrick Fridenson and André Straus, eds., *Le capitalisme français, XIXe–XXe siècles: Blocages et dynamismes d'une croissance* (Paris: Fayard, 1987); and Maurice Lévy-Leboyer and François Bourguignon, *L'économie française du XIXe siècle: Analyse macroéconomique* (Paris: Economica, 1985).

16. "Erlanger avait fait l'emprunt 1863. Son exploit le plus célèbre est la vente au bey, pour un million de francs, de 100 canons rayés défectueux." Jean Bouvier, *Le Crédit Lyonnais de 1863 à 1882* (Paris: SEVPEN, 1961), 495.

17. On British technological supremacy, see Bernard Finn, *Submarine Telegraphy: The Grand Victorian Technology* (London: Science Museum, 1973); Vary T. Coates and Bernard Finn, *A Retrospective Technology Assessment: Submarine Telegraphy; The Transatlantic Cable of 1866* (San Francisco: San Francisco Press, 1979); Georges Demanche, "Les câbles sous-marins à la merci de l'Angleterre," *Revue française de l'étranger et des colonies* (1899): 708–712.

18. M. Audouin, *Les câbles télégraphiques et les intérêts français* (Paris, 1899).

19. H. Bousquet, *La question des câbles sous-marins en France* (Paris, 1895).

20. On Anglo-American rivalry, see Ludwell Denny, *America Conquers Britain: A Record of Economic War* (New York: Alfred A. Knopf, 1930); Pascal Griset, "L'évolution des télécommunications intercontinentales au XXe siècle," *History and Technology* 8 (1992): 231–245; and Pascal Griset, "The Development of Intercontinental Telecommunications in the Twentieth Century," *Flux, Cahiers scientifiques internationaux Réseaux et Territoires* 9 (July–September 1992): 19–31.

21. The affair was ignored in order not to degrade the reputation of the company. As a consequence, Mr. Renard was pursued by the CFCT board of directors. Compte rendu du conseil d'Administration (July 8 and 22, 1885), 1–48, Fonds historiques de la Compagnie française des Câbles télégraphiques, Série F 90, Archives Nationales.

22. M. Bastien, accounting report, in ibid.

23. Such obscuring of financial transactions in French corporations was not unique; see J.-N. Jeanneney, *L'argent caché* (Paris: Seuil, 1983).

24. CFCT, Compte rendu du conseil d'Administration (February 2, 1887).

25. At a cost of between 25,000 and 30,000 francs. See ibid. (March 2, 1887).

26. Ibid. (May 9, 1888). Count Dillon, who had known him at the military academy at Saint-Cyr, determinedly supported the political action of General Boulanger. "Starting in 1886, he appeared in his circle of acquaintances. He was there in particular as organizer of the elections. It is he who advanced the first money for the election of February 26, 1887, for which Boulanger was ineligible. Dillon then launched boulangisme as a business, by a completely new processes of political advertisement. He reproduced millions of cheap photos of the general and millions of song sheets. For the elections, when Boulanger appeared he placed agents in departments, districts, cantons, municipalities." Roman d'Amat and Roger Limouzin-Lamothe, eds., *Dictionnaire de biographie française*, vol. 11 (Paris, 1967), 356. This ability to take up positions of influence was doubtless useful in his "reconquest" of the P.Q. He was convicted after the failure of boulangisme and sentenced to deportation. Pardoned in 1895, he withdrew to Morbihan, where he died in 1922.

27. CFCT, Compte rendu du conseil d'Administration (October 6, 1887). It also marked the return at the head of the company of a man who had been fired some years before.

28. Daniel R. Headrick, "Câbles télégraphiques et rivalités franco-britanniques avant 1914," *Guerres mondiales et conflits contemporains* 166 (April 1992): 133–147.

29. Pascal Griset, "Espace et technologie: Une géopolitique des télécommunications intercontinentales au XIXe siècle," *Sciences de la Société* 35 (May 1995): 88–89.

30. Richard Duboff, "The Telegraph and the Structure of Markets in the United States, 1845–1890," *Research in Economic History* 8 (1983): 253–277.

31. Globally the French network was very weak. The main links—essentially the cables to the colonies—were under foreign control. P. Marcillac, *Les câbles sous-marins: Rapport présenté aux sections du congrès colonial français de 1904.* See Jean-Claude Allain, "L'indépendance câblière de la France au début du XXe siècle," *Guerres mondiales et conflits contemporains* 166 (April 1992): 115–131.

32. Anonymous, "Le service des câbles sous-marins français pendant la guerre," *Annales des P.T.T.* (1921), 236–243; M. Goffin, "Les câbles sous-marins en temps de guerre," *Law Quarterly Review* (London), 15: 145–154.

33. Charles Dupuis, "L'Institut de Droit international et la question des câbles sous-marins en temps de guerre," *Revue générale de Droit international public* 3–4 (1903): 532–555; Charles Dupuis, *Le droit de la guerre maritime d'après les doctrines anglaises contemporaines* (Paris: A. Pedone, 1899).

34. Headrick mentions three reports (1904, 1905, and 1908) that were essential in the preparation of this action. Daniel R. Headrick, *The Invisible Weapon* (New York: Oxford University Press, 1991), 152.

35. Records of the Department of State, 851.73/5, M560, roll 141, National Archives and Records Administration, College Park, Maryland (hereafter NARA).

36. Hoffmann-La Roche to Department of State, April 15, 1915, ibid., 851.73/20.

37. Delcasse to Sharp, Ambassador of the United States, October 1, 1915, ibid., 851.73/35.

38. W. Wilson to R. Lansing, November 18, 1918, ibid., 851.73/119.

39. Standard Oil to Department of State, October 15, 1919, ibid., 851.73/115.

40. The Barter and Commerce Corporation of America to Department of State, March 23, 1921, ibid., 851.73/120.

41. Senate Committee on Interstate Commerce, *A Bill to Prevent the Unauthorized Landing of Submarine Cables*, 92.

42. The memorandum quotes Article 10 of the landing license granted by the French government on July 2, 1895, to the C.F.C.T. Memorandum sent to the Ministère du Commerce, 851.72/13, M560, roll 140, NARA.

43. Ibid. The memorandum quotes Article 6 of the landing contract granted by the French administration on November 30, 1899, to the Anglo American Telegraph Company (which had since rented the cables to Western Union).

44. Ibid.

45. Ibid.

46. S. Pichon, Ministère des Affaires étrangères, to M. Sharp, Ambassador of the United States in Paris, January 31, 1918, ibid., 851.72/19.

47. C. Mackay to F. L. Polk, March 14, 1918, ibid., 851.73/29

48. See for example: Department of State to W. G. Sharp, Ambassador of the United States in Paris, December 21, 1917, ibid., 851.73/13. W. Sharp to Secretary of State, February 2, 1918, ibid., 851.73/19.

49. R. Lansing to C. Mackay, August 30, 1918, ibid., 851.73/45.

50. S. Pichon, Ministère des Affaires étrangères, to the embassy of the United States in Paris, July 30, 1919, 851.73/91.

51. In a normal exchange rate of F 25.22 for £1 and F 5.18 for $1 dollar, the payment by France would have been £299,819 and $1,459,739. In reality, at F 46.95 for £1 and F 12.235 for $1, it was only £161,053 and $618,018. Western Union General Superintendent to Secretary of State, July 23, 1920, Department of State, ibid., 851.73/112. Between February 1918 and August 1920, Western Union evaluated the loss at more than a million dollars. Memorandum, October 25, 1920, Western Union to Department of State, ibid., 851.73/122. For its part, the Commercial Cable Company evaluated its damage for the year 1919 at more than $600,000. C. Mackay to Department of State, July 23, 1920 (letter of 5 pages), ibid., 851.73/114. Memorandum, Commercial Cable Company to Department of State, November 9, 1920, ibid., 851.73/125.

52. The Commercial Cable Company underlines this incoherence in a letter of November 26, 1920, to the Department of State, ibid., 851.73/126.

53. Similar problems arose with Radio France. The sums owed by the administration were the object of rather long negotiations, the company agreeing finally to grant discounts.

54. Pascal Griset, "The Beginnings of the Radio Communications Industry," in *Entrepreneurship and Organization*, ed. M. J. Lynskey and Seiichiro Yonekura (New York: Oxford University Press, 2002), 61–82.

55. W. Rogers, in Senate Committee on Interstate Commerce, *A Bill to Prevent the Unauthorized Landing of Submarine Cables*, 47.

56. Pascal Griset, "Innovation and Radio Industry in Europe during the Interwar Period," in *Innovations in the European Economy between the Wars: Sources and Diffusion of Innovation in Europe between the Wars*, ed. François Caron, Paul Ercker, and Wolfram Fischer (Berlin, New York: Walter de Gruyter, 1995), 37–63.

57. The destruction of Clifden's radio station by the IRA in 1922 demonstrated that no connection passing in transit through Ireland was really secure. W. J. Baker, *A History of the Marconi Company* (London: Methuen, 1971), 226.

58. H. Larose, "Les câbles sous-marins," *L'Illustration économique et financière, numéro spécial* (December 1922), 40–45.

59. It is not useless to indicate that the British retained as a model the contract signed between the Eastern Telegraph Company and the French administration, for the Saigon-Tonkin cable in 1894. Comparing the attitude of the British administration toward the same company on a comparable case, Sir Edward Sassoon, president of the Committee (commission) of the House of Commons on the submarine cables, noted, "The French government has made a good bargain, safeguarding the interest public, and we, here, have made a very bad bargain." He even advised future negotiators to be inspired by clauses stipulated in the French contract, the better to deal with the private companies. E. Sassoon, *Minutes of Evidence Taken before the Inter-Departemental Committee on Cable Communications* (London: Wyman and Sons, 1902), 16.

60. N. Carlton to the president of the American delegation at the international telegraphic conference in Washington, D.C., September 30, 1920, 851.73/138, M560, roll 140, NARA.

61. C. Mackay, in Senate Committee on Interstate Commerce, *A Bill to Prevent the Unauthorized Landing of Submarine Cables*, 264.

62. The AP indicated in addition that its efforts to develop relations with France were without significant results, the French readers not seeming to be interested in the news from America. "I do not think you can artificially set up a distribution of news in countries if they do not want the news." Frederick R. Martin, general manager of the Associated Press, in Senate Committee on Interstate Commerce, *A Bill to Prevent the Unauthorized Landing of Submarine Cables*, 287.

63. This idea anticipates the emergence of the concept of free flow of information, official doctrine of the United States after World War II.

64. N. Carlton, in Senate Committee on Interstate Commerce, *A Bill to Prevent the Unauthorized Landing of Submarine Cables*, 114–115.

65. C. Mackay, in ibid., 279.

66. Let us note moreover that this "British reference" was all the more acceptable for the Americans because all transatlantic transmissions between Great Britain and the United States were operated by American companies.

67. C. Mackay to Secretary of State, January 28, 1920, 841.73/7, M580, roll 65, NARA. The project of rate controls proposed by the British government was in fact very moderate. The loads of the company were totally integrated. Numerous consultations and arbitrations were anticipated.

68. C. Mackay to Department of State, November 26, 1919, 851.73/93, M5560, roll 140, NARA.

69. A memorandum written in 1927 and quoting this document underlines "the Company argued at considerable length." Senate Committee on Interstate Commerce, *A Bill to Prevent the Unauthorized Landing of Submarine Cables*, 263–281.

70. C. Mackay to Department of State, March 9, 1920, 851.73/102, M560, roll 141, NARA.

71. Ibid.

72. C. Mackay to Department of State, April 7, 1921, in ibid., 851.73/171.

73. Its dominant status in the United States sometimes gave its executives the feeling of being closer to the status of a European-style administration. Yet the Commercial executives were also like challengers because of their links with the domestic postal network.

74. Western Union to Secretary of State, March 26, 1920, 851.73/103, M560, roll 140, NARA.

75. Memorandum sent by the Chamber of Commerce of Philadelphia to the Department of Commerce, February 18, 1921, then transmitted in March to the Department of State, 851.73/134, M560, roll 140, NARA.

76. The development of radiotelegraphy gave birth to an even more complex situation. See Pascal Griset, "La société Radio France dans l'entre-deux-guerres," *Histoire, Économie et Société* 1 (1983): 83–109.

77. The position of the Department of State was in any case increasingly delicate. It was indeed more and more difficult for it "to remain neutral" while the two American companies defended opposite interests. The problem of the status of the official messages placed the Department of State in a delicate position because the French company passed on free of charge most messages exchanged by the two governments.

78. "Such a reversal of opinion in the space of two years is rather surprising," concluded this text. Note quoted by Western Union in a letter to the Department of State, April 7, 1923, 851.73/239, M560, roll 141, NARA.

79. The rigor seems to have impressed the American attaché, who noted in a report following a lunch taken together with Lhermite: "He would have no cocktail and no wine—only water—not even milk!" M. G. Jones to F. Colt de Wolf, December 6, 1954, box 5943, file 951.40/12-6-54, RG 59, NARA.

80. N. Carlton to C. Hughes, November 8, 1922, 851.73/233, M560, roll 141, NARA.

81. Pascal Griset, "Les industries de haute technologie en France, 1945–1974," in *Histoire de l'industrie française*, ed. Maurice Lévy-Leboyer (Paris: Larousse, 1997).

82. Pascal Griset, "The Centre National d'Etudes des Télécommunications and the competitiveness of French telephone industry," in *Technological Competitiveness*, ed. William Aspray (New York: IEEE Press, 1993), 176–187; Pascal Griset, "Innover comment? Innover pourquoi? Cinquante ans de recherche au Centre National d'Etudes des Télécommunications," *Cahiers, Télécommunications, Histoire et Société* 1, pt. 2 (1995), 18–35.

83. André Malraux, *Les chênes qu'on abat* (Paris: Gallimard, 1971), 88, 89, 92.

Nations and Networks

9

Strategic and Military Aspects of Submarine Telegraph Cables, 1851–1945

Daniel R. Headrick

The global telegraph cable network of the late nineteenth century was built for commercial purposes, at a time when the major industrial powers were at peace. Only Great Britain had the expertise and capital needed to manufacture and lay cables and the global shipping and trading interests to provide the demand for such a network. From the start, British strategists recognized the value of submarine cables for the defense of Great Britain and the British Empire. Just as cables were a British invention, so was cable strategy.

When France, the United States, and Germany woke up to the strategic value of cables, they found that British firms dominated the field everywhere but across the North Atlantic. In comparison with the British, their thinking about the strategic value of cables was amateurish and ad hoc until just before World War I. Their attempts to build rival networks proved unsuccessful. Instead, they turned to long-distance radio as a means of bypassing the British network entirely.

A paper about cable strategy must perforce include two parts. The first part will deal with the British hegemony in intercontinental telecommunications before 1914, British strategic thinking on the subject of cables, and the reactions of the other powers.[1] The second part will consider the era of the two world wars, showing how strategic planning worked out in actual conflicts.

The Creation of a Global Network, 1851–1898

After a first failed attempt, a submarine telegraph cable was successfully laid across the Channel in 1851. The governments of both Britain and France quickly realized the military potential of the cable. As early as 1854 they had a temporary cable laid across the Black Sea in support of their expedition to the Crimea.[2] The French government was also eager to communicate with its troublesome colony of Algeria. The Mediterranean was much deeper than the Black Sea and had a rocky floor, and the first cables, laid between 1856 1863, broke or were lost overboard.[3]

The failures in the Channel and the Mediterranean were but a prelude to the most spectacular failure of all: the Atlantic cable of 1858. This project was commercial in inspiration, but it had a naval dimension as well. In 1854 the American entrepreneur Cyrus Field began promoting the idea of a cable between Ireland and Newfoundland, the two closest points on either side of the Atlantic. In this, as in most subsequent ventures, naval surveys provided cable entrepreneurs with information about the depth and quality of the ocean floor, a valuable if hidden subsidy.

To lay the cable, Field obtained the loan of two warships, one American and one British. These and other ships failed several times to lay the cable before a successful connection was finally established in August 1858. A month later, the cable suffered a short circuit that ended its useful life. Among the messages conveyed in that month was an order from the British government countermanding a previous order to send a contingent of Canadian soldiers to India. The British Treasury thereby saved £50,000.[4]

The same problems plagued the first cable to India. In 1858 the British government, desperate to communicate with India, where the Sepoy Rebellion threatened to overthrow British rule, offered a generous contract to lay a cable from Egypt to India via the Red Sea. The first section, down the Red Sea to Aden, was laid in 1859 and failed immediately. The ships that laid the cable did not allow sufficient slack, and the cable broke before a single message could be transmitted. The second section, laid in 1860 from Aden to Karachi, also failed.[5]

The failures of the Red Sea and Atlantic cables led the British government to name a special commission to investigate all aspects of submarine telegraphy and set guidelines to be followed in manufacturing and laying cables. With the knowledge gained from the commission's report, entrepreneurs laid a cable to North America in 1865, but it broke two-thirds of the way across. In 1866 they laid a new cable and then went back and repaired the 1865 cable. Both worked well for many years. Cable technology had finally mastered the Atlantic Ocean. In the following decades, cables were laid across all the other seas and oceans.[6] By 1892, British companies owned 155,815 kilometers of cables, or 63.1 percent of the world total.[7]

British Cable Policy, 1866–1891

Until the 1880s, Britain faced no serious rival in submarine telegraphy. Practically all intercontinental telegraphic communications, regardless of origin or destination, traveled over the British lines.[8] Despite their overwhelming superiority in cables, cable ships, and naval power, the British worried about the security of their communications, especially with India. After the failure of the Red Sea cable of 1858, they had to rely on landlines through Prussia, Russia, and Persia, or through Austria and the Ottoman Empire. Not only were these lines unreliable and inefficient, they were vulnerable to the vagaries of British relations with the countries through which they passed. Even the new cables laid from England to India via the Mediterranean and the Red Sea in 1870 required a short landline from Alexandria to Suez, putting Britain's communications with India at the mercy of Egyptian politics.

In 1866 the Select Committee on Communications with India, appointed by the House of Commons, recommended all-British routes wherever possible and, where that was not possible, several alternate routes to India. These became the first principles of British communications strategy. By the 1880s, the naval bases at Gibraltar, Malta, Alexandria, Aden, Bombay, Cape Town, Singapore, Hong Kong, and Halifax were linked to Britain by all-British cables. Still, exceptions like Mauritius and Bermuda made strategists uncomfortable.[9]

In the nineteenth century, nothing stimulated British strategic thinkers as much as a Russian war scare. The scare of 1878 led to the appointment of the Royal Commission on the Defence of British Possessions and Colonies. Its report, issued in 1881, emphasized anew the importance of cables to the trade and defense of the empire. Another scare, in 1885, led the British cabinet to set up a permanent Colonial Defence Committee, with representatives of the Admiralty and the Colonial, Foreign, India, and War offices.[10] This committee reported that British communications were basically safe because cables were hard to find at sea and their landing places were protected by shore guns.[11] Yet it was concerned with Bermuda, the only major British naval base without a telegraphic link to Britain, and the West Indies, which could only communicate with Britain by way of Cuba and the United States. In 1889 the cabinet accepted the committee's recommendation for a Halifax-Bermuda cable at a cost of £8,100 a year.[12] It was the beginning of a trend toward laying ever more costly strategic cables of little commercial value.

Two years later, a committee was appointed to consider the question of telegraphic communication with India in time of war. Its report included the following principles:

> 2. The telegraphic systems of the world are so closely interconnected that the Committee have found it impossible to deal with the question of telegraphic communication with India, without taking into consideration the requirements of the Empire as a whole.

3. The ideal of telegraphic communication for purposes of war would be fulfilled by cables landing on British territory only, following recognized trade routes, and not passing near the naval stations of a possible enemy.[13]

The report concluded that an ideal cable network would be unreasonably expensive, but that the network that Britain already possessed was secure enough for any possible war.

The Spanish-American War and the Neutralization of Cables

The cables that first connected Britain and the United States were greeted with paeans of praise as an invention that would bring peoples together and banish wars. President Buchanan expressed this feeling in the telegram he sent to Queen Victoria via the Atlantic cable of 1858: "May the Atlantic telegraph, under the blessing of heaven, prove to be a bond of perpetual peace and friendship between the kindred nations, and an instrument destined by Divine Providence to diffuse religion, liberty, and law throughout the world." To safeguard this instrument of peace, he wrote: "In this view will not all the nations of Christendom spontaneously unite in the declaration that it shall be for ever neutral, and that its communication shall be held sacred in passing to the place of their destination, even in the midst of hostilities?"[14]

The idea of declaring submarine cables neutral, hence inviolable in times of war, was very tempting. The French government proposed the idea at the international telegraph conference in Paris in 1865. It was taken up again at the Paris conference of 1884. At the insistence of the British government, however, the International Convention for the Protection of Submarine Cables, signed that year and ratified in 1888, contained an article that read: "It is understood that the stipulations of this convention shall in no wise affect the liberty of action of the belligerents." Lord Lyons, the British delegate to the conference, explained: "Her Majesty's Government understands Article XV in this sense, that in time of war, a belligerent, a signatory of the convention, shall be free to act in regard to submarine cables, as if the convention did not exist."[15]

Despite Britain's diplomatic veto, writers, entrepreneurs, and diplomats continued to wonder whether in an actual war, belligerents would respect cables. Would they cut cables out at sea, or only in enemy waters? Would they attack those belonging to a neutral, or those joining two neutrals but carrying information of value to the enemy? The Spanish-American War of 1898 was to answer these questions.

Before that war, Spain could communicate with its colony of Cuba by two American cables between Key West and Havana, a British cable from

Cuba to Jamaica, and a French cable to Haiti and from there to North America. Britain and France were neutral in the war—but were their cables?

When the war began, the United States had two objectives: to stay in touch with its armed forces and to prevent Spain from doing likewise. To communicate with Washington, the American fleet blockading Havana brought up one end of the cable from Key West. Later, when the army invaded Cuba near Santiago, it requisitioned a French-owned cable to Haiti and New York.[16]

Opening communications with the American forces, it turned out, was easier to achieve than cutting those of the enemy. The U.S. Army censored messages on the Key West–Havana cable, but other cables were beyond its reach. Hoping to isolate Cuba from Spain, the U.S. Navy cut a coastal cable, but it failed to find those to Jamaica and Haiti. As Capt. George Squier of the Signal Corps explained, "The searching for deep-sea cables in the high seas in time of war, without an adequate chart of the location of the cable, is a difficult and very doubtful operation."[17]

In the end, the actions of the United States toward neutral cables had no effect on the course of the war. Yet its attempts to interfere with Spanish communications were much commented on by American, French, and German writers, for they demonstrated that cables belonging to neutral powers were just as likely to be seized or cut as the cables of an enemy. Furthermore, cutting cables required both command of the sea and specialized knowledge and equipment. At the time and for years thereafter, only Great Britain possessed such capabilities.

British Cable Strategy and the Boer War, 1892–1901

In the late nineteenth century, Britain found itself involved in increasingly tense incidents with other powers, culminating in a war in South Africa. In the fall of 1898, the cabinet appointed the Committee to Consider the Control of Communications by Submarine Telegraph in Time of War. Its report touched on the same topics as the 1891 report, but at greater length and more adamantly.[18] It prescribed in minute detail what should be done to stock supplies, station cable ships, lay new cables, defend landing sites, and otherwise prepare for war. Like the 1891 report, it recommended the laying of strategic cables that would touch only on British soil, including some very long ones from Gibraltar to South Africa and from there to Australia. Unlike its predecessor, it emphasized the need and disregarded the costs involved.

The committee drew up contingency plans in the event of a war with Russia, Japan, or the United States, but especially France. (Interestingly, the possibility of war with Germany was not even considered.) It showed no confidence in the possibility of neutralizing cables:

> The present Committee are not aware that . . . there has been any advance towards the neutralization of cables in war. The experience of the recent war between Spain and America . . . points to the conclusion that a nation will not hesitate to cut a cable, even when belonging to the subjects of a neutral Power, if it anticipates that any military advantage will be derived from such action. . . . It seems doubtful whether an enemy would be in any way influenced in the direction of fairness and restraint by what this country might do.[19]

Like its predecessors, the committee believed that Britain's best defense was the multiplicity of its cables and its ability to repair them, protected by the power of the Royal Navy.

In one way, the report marked a drastic departure from the defensive approach of the past. Freed of the need to worry about fairness and restraint, "the Committee, after full discussion, have come to the conclusion that we ought to cut an enemy's cables wherever necessary for strategic reasons."[20] Having thus defined a more aggressive strategy, the report went on to list the foreign cables to be cut and the British ones to be interrupted in the event of war with France, with France and Russia, with Japan, and with the United States. It stressed the importance of respecting neutral messages but advocated censorship of telegrams passing through British territory and a discreet scrutiny of messages passing through British-owned cable stations in other countries. It even suggested hiring "an expert in deciphering."

On October 14, two days before the start of the Boer War, Britain imposed censorship at Zanzibar and Aden, forbidding all telegrams in code or cipher except those between foreign governments and their consuls in Africa. This action showed the world that Great Britain was perfectly willing to use its domination of the world cable industry to interfere with the communications of neutral powers in time of war. Since other countries were assumed not to be "influenced by fairness and restraint," it followed that Britain was not either. It was the first lesson in twentieth-century ethics.

Cable Strategies of Other Great Powers, 1898–1914

Between 1898 and 1902, as the neutrality of cables faded from view, French, German, and American strategists awoke to the importance of cables for their security and to their vulnerability to British power. They proposed several remedies. They considered two new ideas: cooperation among the secondary cable powers to free themselves from British control of their communications, and the construction of global wireless networks that would bypass cables entirely.

The Franco-British confrontation at Fashoda in 1898 and British actions during the Boer War brought cables to the attention of the French government. In 1901 the Chamber of Deputies authorized the purchase or laying of cables that would allow France to avoid the British network. The first such cable, connecting Dakar to Pernambuco, Brazil, was purchased from a British company in 1902.[21] The following year, France bought a cable from another British firm, linking Dakar with its colonies along the West African coast. In 1905 it laid a new cable from France to Dakar. By 1912–1913 it operated its own cables to all its West African colonies.[22]

More distant colonies were less well served. To the frustration of French colonialists, Indochina, France's richest possession, could only communicate with France via the British network. To escape that dependency, a cable was laid in 1901 between Indochina and Amoy, in China, the end of a Danish-owned cable and landline across China and Russia. Five years later, another cable joined Indochina to Borneo, where it connected to a German-Dutch cable to Guam and an American cable to the United States. France went to great expense to achieve "a route which, if not entirely French, is at least friendly and neutral."[23] These cables, however, charged higher rates than the British company, and hence they attracted almost no commercial business and consistently lost money. When they broke a few years later, they were abandoned after being deemed not worth repairing.

What were the results of the French desire for cables after 1900? The French overseas network cost between six and eight million francs (£240,000–320,000) a year in subsidies. Clearly these cables were not laid for commercial reasons. Their ostensible purpose was strategic—to avoid British censorship in the event of a war. Although that fear evaporated when France and Britain signed the Entente Cordiale in 1904, the cable-laying momentum continued, impelled by the vested interests it had created and the continued efforts of the cable lobby.[24] Thus, the cables were not really strategic either; rather, they were political, designed to place France among the great powers in the realm of global communications.

Like France, Germany discovered the drawbacks to its dependence on British cables in the 1890s. Its colonies, acquired late, were too distant and poor to warrant the enormous investment in a separate cable network. Furthermore, its geographical position long frustrated any telegraphic plans. When the German government at last decided to build its own network in 1894, it realized it needed allies.

Beginning in 1900, political pressure and economic demand combined to trigger a cable-laying boom. The demand came from Germany's growing trade, especially with the Americas. Even had there been no political motives, this trade would have justified cables to North and South America. What politics did, as in France, was to divert cable investments to less profitable routes to Africa and in the Pacific, where German business interests were marginal at best.

As in France, it was censorship during the Boer War that triggered the anti-British indignation that led to such investments. The first investment, an annual subsidy of 1.4 million Reichsmark (£67,000), went toward a new cable to the United States via the Azores.[25] It avoided British territory but required the approval of Portugal, an ally of Britain.

After the North Atlantic, the next objective of Germany's cable policy was the Ottoman Empire, where Germany had growing interests. Before 1899, communications with Turkey went either by landline across Russia or by British cables in the Mediterranean, neither one politically acceptable. In 1899 a German firm obtained the right to construct a landline across Hungary and Romania, but was blocked until 1905 from laying a cable to Constantinople by British pressure on the Ottoman government.[26] Meanwhile, Germany was building a railroad and telegraph line from Constantinople to Baghdad, which could easily have been prolonged by a telegraph line to the Persian Gulf. From there the German government intended to lay a cable across the Indian Ocean to Portuguese Goa and Dutch Sumatra. In 1913, however, British Foreign Secretary Sir Edward Grey blocked German access to the Indian Ocean on the grounds that "Germany has no great commercial need for a cable such as that described . . . the object of their request is political."[27]

Like French Indochina, Germany's colony and naval base at Kiaochow Bay on the Chinese coast could communicate either via British cables or via Russian landlines, neither of them politically and strategically satisfactory. For its Pacific colonies—the Carolinas, Marianas, Palau, and part of Samoa—it looked for assistance to the Dutch and the Americans. Throughout the nineteenth century, the Netherlands had lived in the shadow of Britain, with some misgivings, especially during the Boer War. In 1900, when a German cable entrepreneur approached the Dutch Telegraph Administration about a joint cable connecting the Netherlands East Indies with the German colonies in the Pacific and eventually with an American Pacific cable, he found a warm reception. In 1903, when the American cable reached the Philippines, a new Dutch-German company laid cables linking the German island of Yap with American Guam, Kiaochow Bay, and the Netherlands East Indies. The German and Dutch colonies could now communicate with their metropoles by non-British lines.[28] That businesses continued to use the less costly British lines was beside the point. As Richard Hennig, a German expert on communications, explained, "The new lines do not meet any requirement of business nor any requirement of peace, their purpose is much more an eminently strategic one, and just laying them constitutes a move against British policy."[29]

In the South Atlantic, the Germans faced the same dilemma as in the Pacific, namely, the impossibility of building their own cable network. Spain and Liberia were willing to grant concessions, but on the Brazilian side, if they wished to avoid the British, they had no choice but to turn to the South American Telegraph Company, owned by the French government. In exchange for a share in this company's concession, Germany helped France complete its West African network.[30]

In their desire to escape British control over their communications, France and Germany were thus thrown together into a marriage of convenience. Commentators on both sides noted that the cable interests of the two nations ran parallel. French public opinion, however, was shocked, for in other areas, France and Britain were allies, while Franco-German relations were becoming increasingly hostile. In other words, the cable policies of the two nations were politically outdated. As for their strategic value, Britain's commercial and naval dominance of the sea ensured that it was nil.

American companies were second only to the British in acquiring cables. By 1911 Western Union and Commercial Cable Company controlled the entire North Atlantic cable network. American companies also owned cables in the Caribbean and Central America and along the west coast of South America. Yet the U.S. government had no cable policy and neither aided nor interfered with commercial ventures.

The Pacific was far wider than the Atlantic and had only a fraction of the trade. Before 1898 American businessmen had several times proposed to lay a cable from California to Hawaii and East Asia, but they were unable to obtain a subsidy from Congress.[31] In 1898, however, the United States found itself with a Pacific empire stretching from Hawaii to Guam and to the Philippines, making a Pacific cable a strategic necessity. In messages to Congress, Presidents McKinley and Roosevelt pleaded for money to lay a cable.[32] Rear Admiral Bradford of the U.S. Navy and Captain Squier of the U.S. Army Signal Corps also lobbied Congress, as did several cable companies. But Congress was not yet convinced that an empire cost money to maintain and again refused to grant any subsidy.[33]

In August 1901 the Pacific cable issue, which had been tossed around for twenty-five years, was suddenly resolved when Clarence Mackay, president of the Commercial Cable Company and Postal Telegraph Cable Company, offered to lay a cable to the Philippines without subsidies or landing licenses, on the grounds that the Pacific Ocean was a "navigable water of the United States." His new enterprise, the Commercial Pacific Cable Company, laid a cable to Manila in 1903, prolonging it to China and Japan in 1906.[34]

BRITISH CABLE STRATEGY, 1902–1914

The shock of Britain's isolation in the Boer War focused the attention of the British government on the strategic value of its cables. One question that immediately came to the fore was rounding out the British network by laying a cable across the Pacific Ocean. Such a cable had often been proposed from the 1870s on but had always run into the strident opposition of the Eastern group, which enjoyed a lucrative monopoly on communications between Australia, New Zealand, and the rest of the world. In 1901, reacting to the international tensions since 1898, the British parliament created the Pacific Cable Board, a public entity jointly owned by Britain, Canada, New Zealand, and the Australian colonies, which laid a cable in 1902. From that moment on, Britain's network encircled the globe.[35]

In the aftermath of the Boer War, the armed forces and the Colonial Defence Committee developed what historian Paul Kennedy called a "virtual fetish" for all-British cables.[36] In response to their demands, the British network was rounded out with new cables in the South Atlantic and Indian oceans. While they duplicated existing lines, they would provide alternative routes to India in the event that Britain lost control of the Mediterranean Sea. These cables, of little commercial value, proved to be strategically vital in the wars to come.

After 1902 Britain possessed not only most of the world's commercial cables but also a set of strategic cables that made its communication with its major colonies and naval bases practically invulnerable. But having secure communications was not enough. In a world where war seemed likely, strategists made ever more elaborate contingency plans and worried about defensive and offensive actions in minute detail. In its report dated March 26, 1902, the Inter-Departmental Committee on Cable Communication declared itself satisfied with the security of Britain's global communications and turned its attention to offensive operations.[37] In 1904 the Admiralty issued its "Memorandum on the Protection of British Submarine Cables, and on the Destruction of an Enemy's Cables."[38] That same year, the War Office prepared a document entitled "Censorship of Cables in Time of War." Noting that the International Telegraph Convention of St. Petersburg required that other governments be notified before any telegraph service was suspended "except where the advice might be dangerous to the security of the State," the War Office noted that "wherever there was a suspension of service under that Article it would be held dangerous for the security of the State to advise the office of origin, and therefore no advice should be sent."[39]

By 1908 British strategists realized that flaunting their power over the communications of neutral countries, as they had done in the Boer War, was counterproductive. The report of the "Inter-Departmental Committee appointed to draw up instructions for establishing a secret censorship of submarine cables at certain foreign stations in time of grave emergency" noted:

> During a period of strained relations with a foreign Power, when war, though not inevitable, may break out at any moment, it may become necessary, at certain stations abroad, to take measures to prevent the disclosure of information regarding naval or military preparations and movements. The inauguration of the scheme of open cable censorship arranged to be put in force in time of war might in such circumstances precipitate hostilities; it is therefore necessary to resort to some form of secret censorship.[40]

The last major report on cable strategy before World War I was issued by the Standing Subcommittee of the Committee of Imperial Defence in December 1911 under the title "Submarine Cable Communications in Time of War."[41] It noted with satisfaction that "the dependence of the United Kingdom on cable stations situated upon foreign territory for the transmission of telegrams has been generally eliminated." Turning to the offense, the subcommittee gave detailed instructions on cutting Germany's intercontinental communications: "Generally speaking, if France and Russia were in alliance with this country, it would be possible to isolate Germany from practically the whole world, outside Europe, by cutting the cables to the Azores, Tenerife, and Vigo, and the three cable landings on Yap Island."

By 1911 Great Britain had finally achieved the ultimate goal of communications strategy. It had the ability to cut enemies off from the rest of the world and secretly to censor or scrutinize the messages of neutrals, while its own cable network was invulnerable to attack, interference, or espionage.

Cables in World War I

In contrast to the flagrant unreadiness of Britain's army in 1914, and even to the uncertainties faced by the Royal Navy, Britain's planning in the area of cable communications and strategy was practically flawless.[42] Within hours of the expiration of Britain's ultimatum to Germany on August 4, 1914, the cable ship *Telconia* cut the five German cables to the Americas and Africa. Within days, France and Russia had severed communications between Central Europe and the rest of the world. German cables in the Atlantic and Pacific oceans and off the coast of China were also shut down or cut in the first weeks of the war. By early 1915 all of Germany's cables were either cut or in Entente hands.[43]

Germany also attempted to harm the communications of the Entente, but it was a more daunting task. As soon as the war began, Germany closed Britain's landline connections to India. In September 1914 it cut the Baltic cables between Russia and its western allies. In November, the Ottoman Empire severed the Black Sea cables. Two months later, Britain laid a new cable to Russia via the Arctic Ocean, restoring secure communications. Low-priority communications with India and East Asia were often delayed, but official messages were never interrupted.[44]

More dramatic, but less important from a communications standpoint, were German attacks on distant cable stations. In September 1914 a landing party from the German cruiser *Nürnberg* seized the cable station on Fanning Island and damaged the Pacific cable, but it was soon repaired. In a more spectacular raid, a party from the German cruiser *Emden* landed on Cocos Island in the Indian Ocean, an important cable station between Australia, Ceylon, and Singapore. The Germans destroyed some equipment and cut some dummy cables, but they missed the main cable to Singapore. Alerted by telegram, the Australian cruiser *Sydney* rushed to the scene and sank the *Emden*, but the landing party managed to escape. Although the raid made an exciting story, the damage it caused was repaired within twelve hours after the Germans left.[45] After that, the Germans sent submarines to cut cables. They succeeded several times in the North Sea and in the shallow waters off the coasts of Portugal, Newfoundland, and the United States, but the breaks were soon repaired.[46]

By 1915 the Central Powers relied on the goodwill of Sweden or the United States to smuggle out their messages by radio or on British cables. Both routes, however, were vulnerable to British code breakers.[47]

The overseas communications of the Allied Powers, in contrast, not only survived almost intact but increased enormously, stimulated by their governments' demand for communications, especially after the United States entered the war in 1917. By 1918 the British government alone was cabling 30,000 words a day, ten times as many as in 1913.[48] Press and private telegrams were often delayed for days, sometimes weeks, but official messages always got through on time and safe from German espionage.

To satisfy the booming demand, France, Britain, and the United States built powerful radio stations, a topic that lies outside the scope of this article.

Long-distance radio, however, was both unreliable and insecure, and cables were used to their maximum capacity. To increase the capacity of their network, the Allies used the captured German cables whenever possible. By relaying captured German cables, the three allies increased their cable networks considerably and, in doing so, created subjects for postwar quarrels.[49]

Besides trying to cut off their enemies' channels of communication, the belligerents strove to deprive their enemies of useful intelligence by supervising the export of information. Beginning on August 2, 1914, censors were stationed at all British cable stations around the world, following the plans devised before the war. In places where customers had a choice of cables owned by another country, overt censorship was replaced by the secret scrutiny of telegrams.[50] Britain used its censorship not only to prevent military information from reaching its enemies but also to control world trade to its advantage. Neutral shipping became an adjunct to the allied war effort by the simple expedient of denying access to the cable network to any ship, bank, or business deemed to be engaged in "war contraband." From October 1917 to November 1918, Britain even cut off the Netherlands and its colonies, though they were neutrals in the war, until the Dutch ceased allowing German shipments of sand and gravel to the Western Front. Moreover, they deliberately held back commercial telegrams to Europe up to 48 hours, despite protests from France, Italy, and the British public. Before major military operations, the delay might be a long as a week, to ensure that any spy messages that might have escaped the censors would arrive too late to be of use to the enemy.[51]

In addition to preventing information from reaching the enemy, cable censorship was also instrumental in controlling the flow of news from the belligerents to the world outside Europe and to enlist the aid and sympathy of neutrals, especially the United States. According to one journalist, three quarters of the dispatches from American correspondents in Central Europe were banned by the British censors.[52]

In the final analysis, cable censorship was of minor value in catching spies and obtaining information about the enemy. Instead, its real value lay in halting trade between the Central Powers and the rest of the world and in preparing American public opinion to enter the war on the side of the Entente.

Cable and Strategy between the Wars

When the leaders of the victorious Allies assembled in Paris in 1919, they were concerned with maximizing their own country's means of communications, even at the expense of global efficiency. Britain, France, and Japan did so brazenly, while the United States hid its national interest behind a cloud of high-minded rhetoric.

A major bone of contention was the German cables severed in 1914. Of the two German North Atlantic cables, one now connected Britain and Newfoundland, and the other linked France and the United States. Germany's South Atlantic cables had been taken over by France, while its Pacific cables, along with the important cable island of Yap, were held by Japan.[53]

The Americans saw communications as a much broader issue than just the matter of the German cables. They denounced the British stranglehold on communications based on exclusive landing rights and Britain's habit of using this control to favor British trade at the expense of others; it was the same argument that the French and Germans had used twenty years before. But they added a uniquely Wilsonian touch, advocating open and free communications under international control, which would have transformed telecommunications from an instrument of national power to a global public utility.[54]

To the British, French, and Japanese, the cables and the island of Yap were prizes of war that they had no desire to relinquish, let alone return to Germany. The British Admiralty summed up the British side in a letter of July 5, 1919, to the Imperial Communications Committee:

> The ex-German cables are prizes of war, and their capture was a legitimate application of the use of Sea Power. . . . No reason is seen why we should agree to any international control by which our present strong position would be weakened or our communications hampered in any way.[55]

The British saw themselves not as monopolists but as victims of big American corporations. Faced with the resolve of France, Britain, and Japan, Wilson abandoned his attempts to restructure the world's cable network and postponed the issue of the German cables to a later time. Although the question of the former German cables was revisited in 1920 and in 1921, all the United States gained was the right to land cables and operate radio and telegraph stations on the Japanese-held island of Yap.[56]

For a few short years in the 1920s, the cable business thrived with the booming demand for intercontinental communications. In 1923 the world cable network was over twice as long as it had been in 1892. Old cables were improved by installing regenerative repeaters at each cable station and automatic printers at each end. New cables across the Atlantic and Pacific oceans were several times faster than the prewar cables.[57]

This rosy picture suddenly darkened with the introduction of shortwave "beam" radio by Guglielmo Marconi between 1924 and 1927. The new technology offered such fast and inexpensive service that it soon captured up to 70 percent of the intercontinental traffic and threatened to drive the cable companies out of business. This had serious strategic implications.

In 1927, when shortwave radios had proliferated throughout the world, even in the Royal Navy, the Committee of Imperial Defence convened a special Sub-Committee on Competition between "Beam" Wireless and Cable Services. In its report dated December 8, 1927, the subcommittee concluded that radio was vulnerable to code breaking, and therefore, "We have assumed that all Governmental communications of a secret or confidential nature will be transmitted by cable, as well as any communications of a quasi-commercial character which might, if intercepted, convey information of use to the enemy."[58]

At the very same time, the chairman of Eastern announced that his company, rather than go bankrupt, was thinking of selling its cables to foreign countries, distributing its cash reserves to its shareholders, and going out of business. For the first time, Britain's strategic needs conflicted with the financial interests of the cable business. Faced with this threat, the British government merged all public and private cable and radio enterprises into a new enterprise, Imperial and International Communications, Ltd., under the control of the Imperial Communications Advisory Committee. The goal of this committee was to prevent the closing of money-losing cables that might be needed for defense, even at the expense of the company's profits.[59]

As the Depression worsened, the conflict between commercial and strategic interests intensified. The Imperial Communications Committee called a Strategic Cables Sub-Committee to investigate the pleas of the company for subsidies. Its report, dated March 1933, reiterated the need for cables, though it allowed the company to put certain cables on a "care and maintenance" footing, ready to be brought back on line in a month's time.

Other nations reacted differently to the conflict between cables and radio. France followed Britain's example by merging its communications interests.[60] The United States government forbade such a merger on antitrust grounds.[61] As radio proliferated and military planners foresaw wars of movement, attention shifted away from cables and toward the development of the cipher machines needed for secure radio communications. From their experience in World War I, the German military drew the conclusion that cables were of little value in war and placed their trust in a cipher machine—the famous Enigma.

Cables in World War II

From the standpoint of cables, World War II was in some ways a repeat of World War I. Immediately upon entering the war, Britain cut Germany's Atlantic cables and imposed censorship, secret scrutiny, or deliberate delays on all traffic flowing through its cables and stations. Yet attacks on cables were less one-sided than they had been in the First World War. Though the British cut the Italian cables in the Mediterranean Sea as soon as Italy entered the war, the Italian navy was able to cut most of the British cables there as well. The British base on Malta was left with only two cables to Egypt and its cable communications with Britain had to go around Africa or, for a time, via Australia and Canada.[62]

Similarly, in the days after the attack on Pearl Harbor, the Japanese seized or cut almost all the British, French, Dutch, and American cables in East Asian waters, but they showed no interest in operating them.[63] In March 1942 a Japanese warship shelled Cocos Island, damaging the cable station and other buildings; but the cables survived and the flow of messages was not interrupted.

From a communications standpoint, however, the situation had changed radically since World War I, for radio was now ubiquitous and communications involved a mix of cables, radio, and encryption. Radio traffic, many times more intense than in the previous war, was subject to traffic analysis, direction finding, and cryptanalysis by German as well as British military intelligence, and later by Americans as well.

Perhaps the most famous incident involving cables occurred just before the Battle of Midway in June 1942. American naval intelligence intercepted a radio message announcing that the Japanese fleet was preparing to attack a place known only as "AF." To determine what "AF" stood for, American intelligence officers in Hawaii sent a message by cable to the garrison at Midway, instructing them to send a plaintext radio message stating that their water distilling plant had broken down. Japanese messages, decrypted by U.S. naval intelligence, repeated the news that the distilling plant at "AF" had broken down. The U.S. Navy thereby learned the objectives of the Japanese fleet, a decisive factor in the upcoming battle.[64]

Once the United States was committed to fighting in North Africa and Europe, transatlantic communications became a matter of strategic urgency. At first, the Americans were inclined to use radio, but they soon learned to appreciate the value of cables. When the U.S. First Army landed in Algiers in late 1942, one of the severed Gibraltar-Malta cables was diverted to that city. In 1943, as the Allied forces advanced from North Africa to Sicily and Italy, cables were laid to connect them to the British network.

To avoid British scrutiny or delays without laying cables of their own, the Americans leased dedicated channels on existing cables. In late 1943 U.S.-controlled transatlantic channels proliferated. In 1944, in preparation for the Normandy invasion, David Sarnoff, General Eisenhower's special assistant in charge of communications, leased channels on both cable and radio circuits. By D-Day, Allied communications with their armies on the European continent were able to handle over a million words a day, more than enough for all their needs.[65]

CONCLUSION

For almost a century, global communications were dominated by a single nation. The military aspects of intercontinental communications were inseparable from their commercial aspects, as submarine cables served both purposes. Great Britain's head start in the manufacture, laying, and operation of cables gave it not only commercial but also strategic benefits. British strategists, aware of the advantage that their country's dominance of the cable business provided, prepared meticulous plans to use that dominance to maximum effect in wartime. Faced with this dominance, other cable-owning nations were in no position to develop coherent cable strategies.

Britain's dominance of cables paid off in both world wars. In World War I, Britain effectively cut the Central Powers off from the world outside Europe, especially from the United States, and thereby contributed to the entry of the United States into the war and to the Allied victory. In World War II, thanks to cables, the Anglo-American alliance enjoyed more secure communications than the Axis powers, which were forced to rely on radio and encryption machines. Cables and cable strategy played a vital, if largely unappreciated, role in the outcome of both world wars.

Communication involves both interactions between human beings and the media that carry those interactions. Periodically, newly invented means of communication promise to change not only the way in which people interact but also the goals and effects of their communication. In 1858 President Buchanan wrote feelingly about the era of "perpetual peace and friendship" that submarine cables had introduced. A century and a half later, Jaron Lanier, inventor of "tele-immersion," the next stage in communication, had this to say: "Because tele-immersion is fundamentally a tool to help people connect better, the question is really about how optimistic one should be about human nature. I believe that communications technologies increase the opportunities for empathy and thus for moral behavior."[66]

President Buchanan may be forgiven his innocence, for he lived at a time of widespread faith in progress. Two world wars and countless minor conflicts later, we should know better. Improved means of communication serve the gods of war as readily as the spirits of peace and empathy.

Notes

1. Paul M. Kennedy, "Imperial Cable Communications and Strategy, 1870–1914," *English Historical Review* 86 (1971): 728–752; Daniel R. Headrick, *The Invisible Weapon: Telecommunications and International Politics, 1851–1945* (New York: Oxford University Press, 1991), chaps. 2–6.

2. "Correspondence Respecting the Establishment of Telegraphic Communications in the Mediterranean and with India," *Parliamentary Papers, 1857–8* (2406), lx. See also Willoughby Smith, *The Rise and Extension of Submarine Telegraphy* (London: J. S. Virtue, 1891), 40–42.

3. *Exposé du développement des services postaux, télégraphiques et téléphoniques en Algérie depuis la conquête* (Algiers: J. Carbonel, 1930), 49–50; Charles Bright, *Submarine Telegraphs: Their History, Construction, and Working* (London: Crosby Lockwood, 1891), 61; K. R. Haigh, *Cableships and Submarine Cables* (London: Adlard Coles, 1968), 302–304; Georg Siemens, *History of the House of Siemens*, vol. 1, *1847–1914* (Freiburg: Karl Alber, 1957), 59, 108–110.

4. George Saward, *Deep Sea Telegraphs: Their Past History and Future Progress* (London: Mechanics Magazine, 1861), 45; Vary T. Coates and Bernard Finn, *A Retrospective Technology Assessment: Submarine Telegraphy; The Transatlantic Cable of 1866* (San Francisco: San Francisco Press, 1979), 12–13.

5. "Correspondence Respecting the Establishment of Telegraphic Communications" and "Further Correspondence Respecting the Establishment of Telegraphic Communications in the Mediterranean and with India," *Parliamentary Papers, 1860* (2605), lxii; "History of Telegraph Communications with India (1858–1872) and an account of Joint Purse from 1874" (1897), 83/56, pp. 3–23, General Post Office Archives, London. See also Halford Lancaster Hoskins, *British Routes to India* (New York: Longmans, Green, 1928), 373–397.

6. Coates and Finn, *A Retrospective Technology Assessment*, 21–25. See also George P. Oslin, *The Story of Telecommunications* (Macon, Ga.: Mercer University Press, 1992), 173.

7. U.S. Department of the Navy, Bureau of Navigation, Hydrographic Office, *Submarine Cables* (Washington, D.C.: G.P.O., 1892), 41–59.

8. Peter J. Hugill, *Global Communications since 1844: Geopolitics and Technology* (Baltimore: Johns Hopkins University Press, 1999), 39–45.

9. "Report of the Select Committee Appointed to Inquire into the Practical Working of the Present System of Telegraphic and Postal Communications between this Country and India," *Parliamentary Papers, 1866* (428), ix; Vice-Admiral Arthur R. Hezlet, *The Electron and Sea Power* (London: P. Davies, 1975), 9–10.

10. Robert J. Cain, "Telegraph Cables in the British Empire 1850–1900" (PhD diss., Duke University, 1971), 207–211 and 228.

11. "Protection of Telegraph Cables in Time of War," memorandum by G. S. Clarke, secretary of the Colonial Defence Committee, August 5, 1885, Cabinet Papers, CAB 8/1, No. 12M, Public Record Office (hereafter PRO). See also Sir James Anderson, *Cables in Time of War* (London, 1886), 9; Anderson was one of the directors of the Eastern Telegraph Company.

12. Memorandum of May 12, 1885, on the Bermuda cable, Cabinet Papers, CAB 8/1, No. 4M, PRO. Cain, "Telegraph Cables in the British Empire," 212–213.

13. "Report of a Committee appointed to Consider the Question of Telegraphic Communication with India in Time of War, Juland Danvers, Chairman," March 19, 1891, Cabinet Papers, CAB 18/16/2, PRO.

14. S.A. Garnham and Robert L. Hadfield, *The Submarine Cable: The Story of the Submarine Telegraph Cable from Its Inception Down to Modern Times* (London: S. Low, Marston & Co., 1935), 31.

15. Captain George O. Squier, U.S. Army Signal Corps, "The Influence of Submarine Cables upon Military and Naval Supremacy," *National Geographic Magazine* 12, no. 1 (January 1901): 8; George Grafton Wilson, *Submarine Telegraphic Cables in Their International Relations: Lectures Delivered at Naval War College, August 1901* (Washington, D.C.: G.P.O., 1901), 12–13; Keith Clark, *International Communications: The American Attitude* (New York: Columbia University Press, 1931), 156–161.

16. On the French cables to and in the Caribbean, see Pascal Griset, *Entreprise, technologie et souveraineté: Les télécommunications transatlantiques de la France (XIXe-XXe siècles)* (Paris: Editions Rive Droite, 1996), 104–105.

17. Squier, "The Influence of Submarine Cables," 11.

18. "Report of Committee Appointed to Consider the Control of Communications by Submarine Telegraph in Time of War (October 22, 1898)," Cabinet Papers, CAB 18/16/4, PRO.

19. CAB 18/16/4 (1898), 13, no. 4, PRO.

20. Ibid., no. 3.

21. Charles Lesage, *La rivalité anglo-germanique: Les câbles sous-marins allemands* (Paris: Plon-Nourrit, 1915), 122–154.

22. "The West African Telegraph Company, Ltd., Agreements No. 2," box 1096, Cable and Wireless Archives, Porthcurno, United Kingdom; "Arrangement Relating to the Sale to the French Government of the Cables Serving the Stations of Conakry, Grand Bassam, Kotonou and Libreville," February 10, 1902, 924/243, Cable and Wireless Archives. See also C. Bouerat, "Les débuts du Service des Postes et Télégraphes en Côte d'Ivoire (1880–1905)," *Bulletin de la Société internationale d'histoire postale* 19–20 (1972): 11–93; and *Les postes et télégraphes en Afrique occidentale* (Corbeil: E. Crété, 1907), 99–100.

23. "Câbles en Indochine," no. FO90 bis 4460, PB332, Archives of the Ministry of Posts and Telecommunications, Paris. See also A. Berbain, *Note sur le service postal, télégraphique et téléphonique de l'Indochine* (Hanoi-Haiphong: Imp. d'Extrême-Orient, 1923), 6.

24. Léon Jacob, "Les intérêts français et les relations télégraphiques internationales," *Bureau des questions diplomatiques et coloniales* (1912); "Câbles sous-marins et défense nationale," *Revue de Paris*, December 10, 1910, 877–903. See also Jean-Claude Allain, *Agadir 1911* (Paris: Université de Paris I, 1976), 133–135.

25. Artur Kunert, *Geschichte der deutschen Fernmeldekabel. II. Telegraphen-Seekabel* (Cologne-Mülheim: Karl Glitscher, 1962), 246.

26. Ibid., 292–301; Lesage, *La rivalité anglo-germanique*, 237–239.

27. "Germany: Telegraphic Communication with China, 1912–1913," Cabinet Papers, CAB 17/75, PRO.

28. R. Hennig, "Die deutsche-niederländische Telegraphenallianz im Fernen Osten," *Grenzboten* 65 (April-June 1906): 289–293; Günther Meyer, "German Interests and Policy in the Netherlands East Indies and Malaya, 1870–1914," in *Germany in the Pacific and Far East, 1870–1914* (St. Lucia, Queensland: University of Queensland Press, 1977), 77; Jorma Ahvenainen, *The Far Eastern Telegraphs: The History of Telegraphic Communications between the Far East, Europe and America before the First World War* (Helsinki: Suomalinen Tiedeakatemia, 1981), 175–184.

29. Hennig, "Die deutsche-niederländische Telegraphenallianz im Fernen Osten," 291–292.

30. Kunert, *Geschichte der deutschen Fernmeldekabel*, 261–274, 345; Lesage, *La rivalité anglo-germanique*, 101–122, 155–160, 247–250.

31. Leslie Bennett Tribolet, *The International Aspects of Electrical Communications in the Pacific Area* (Baltimore: Johns Hopkins University Press, 1929), 156–158, 168–179; W. D. Alexander, "The Story of the Trans-Pacific Cable," *Hawaii Historical Journal* 18 (January 1911): 50–65.

32. "An American Pacific Cable. A Paper Presented at the One Hundred and Thirty-eighth Meeting of the American Institute of Electrical Engineers, New York, December 27, 1899, by George Owen Squier," Document no. 88, in U.S. Senate, 56th Cong., 1st sess.

33. Documents 85–89, referred to the Committee on Naval Affairs, January 19, 1900, in U.S. Senate, 56th Cong., 1st sess., esp. Document 89, "A United States Government Pacific Cable" by George O. Squier. See also Ahvenainen, *The Far Eastern Telegraphs*, 166–170; Alexander, "The Story of the Trans-Pacific Cable," 68–80; and Tribolet, *The International Aspects of Electrical Communications in the Pacific Area*, 181.

34. "Pacific Cable," no. 1558/6, Cable and Wireless Archives; Ahvenainen, *The Far Eastern Telegraphs*, 172–173, 187–189; Tribolet, *The International Aspects of Electrical Communications in the Pacific Area*, 182–191, 242; Alexander, "The Story of the Trans-Pacific Cable," 69–71.

35. On the British Empire's Pacific cable, see George Johnson, ed., *The All-Red Line: The Annals and Aims of the Pacific Cable Project* (London: Stanford, 1903); Graham M. Thompson, "Sandford Fleming and the Pacific Cable: The Institutional Politics of Nineteenth-Century Telecommunications," *Canadian Journal of Communications* 15 (1990): 64–75; and Hugill, *Global Communications since 1844*, 45–46.

36. Kennedy, "Imperial Cable Communications and Strategy," 733.

37. "Inter-Departmental Committee on Cable Communications. First Report. Balfour of Burleigh, chairman," August 8, 1901, and "Second Report," March 26, 1902, Cabinet Papers, CAB 18/16/5, PRO; a censored version appeared in *Parliamentary Papers, 1902* (1056), xi.

38. Draft sent to the Foreign Office, August 9, 1904, FO 83/2196, nos. 294–297, PRO.

39. "Censorship of Submarine Cables in Time of War. Action to Be Taken by Her Majesty's Government and by Different Government Departments concerned. Prepared by the General Staff, War Office," FO 83/2196, nos. 387–427, PRO; a similar document appears in Cabinet Papers, CAB 17/92, PRO.

40. Cabinet Papers, CAB 17/92, PRO.

41. "Standing Sub-Committee of the Committee of Imperial Defence. Submarine Cable Communications in Time of War. Report," December 14, 1911, Cabinet Papers, CAB 16/14, PRO.

42. On the role of cables in World War I, see Hugill, *Global Communications since 1844*, 46–48; Griset, *Entreprise, technologie et souveraineté*, 261–279; and Headrick, *The Invisible Weapon*, chap. 8.

43. "1914–1918 War Crisis. Correspondence concerning & instructions to stations re censorship and operations during the war," 1490, Cable and Wireless Archives. See also Patrick Beesly, *Room 40: British Naval Intelligence, 1914–1918* (London: Hamilton, 1982), 2; and Kunert, *Geschichte der deutschen Fernmeldekabel*, 349–351.

44. "Telegraph Communications in Time of War: Direct Wireless Telegraph Communications with Russia," March 7, 1913, Cabinet Papers, CAB 4/5/2/173, PRO; G. L. Lawford and L. R. Nicholson, *The Telcon Story, 1850–1950* (London: Telegraph Construction and Maintenance Co., 1950), 90; Hezlet, *The Electron and Sea Power*, 85; Kunert, *Geschichte der deutschen Fernmeldekabel*, 359–360, 709.

45. Gerald M. Garratt, *One Hundred Years of Submarine Cables* (London: HMSO, 1950), 33; Kenneth C. Baglehole, *A Century of Service: A Brief History of Cable and Wireless Ltd. 1868–1898* (London: Bournehall Press, 1979), 24–25; Hugh Barty-King, *Girdle Round the Earth: The Story of Cable and Wireless and Its Predecessors to Mark the Group's Jubilee, 1929–1979* (London: Heinemann, 1979), 164; Garnham and Hadfield, *The Submarine Cable*, 188–191.

46. Jonathan Reed Winkler, *Nexus: Strategic Communications and American Security in World War I* (Cambridge, Mass.: Harvard University Press, 2008), 117–119; F. Birch and W. F. Clarke, *A Contribution to the History of German Naval Warfare 1914–1918* (1920), 1:649;

Kunert, *Geschichte der deutschen Fernmeldekabel*, 702–712; R. H. Gibson and Maurice Prendergast, *The German Submarine War, 1914–1918* (London: Constable, 1931), 307; Robert M. Grant, *U-Boat Intelligence, 1914–1918* (London: Putnam, 1969), 151–152.

47 Barbara Tuchman, *The Zimmermann Telegram* (New York:Viking, 1958).

48. Barty-King, *Girdle Round the Earth*, 175.

49. Garnham and Hadfield, *The Submarine Cable*, 177–179; Kunert, *Geschichte der deutschen Fernmeldekabel*, 349–358; Barty-King, *Girdle Round the Earth*, 172–175.

50. The prewar censorship plans can be found in "Censorship of Submarine Cables in Time of War" (1904), Cabinet Papers, CAB 17/92, PRO; "Censorship of Submarine Cables in Time of War. Action to be taken by Her Majesty's Government and by Different Government Departments Concerned. Prepared by the General Staff, War Office," May 30, 1905, FO 83/2196, nos. 387–427, PRO; "Report of an Inter-departmental Committee Appointed to Draw up Instructions for Establishing a Secret Censorship of Submarine Cables at Certain Foreign Stations in Time of Grave Emergency," 1908, Cabinet Papers, CAB 17/92, PRO; and "Regulations for Censorship of Submarine Communications and Frontier Land Lines throughout the British Empire and Radio-Telegraphy in the Overseas Possessions and Protectorates," 1913, in WO 33/610. For censorship during the war, see *Report on Cable Censorship during the Great War (1914–1919)* prepared by Colonel Arthur Browne, chief cable censor, in DEFE 1/130, 65–66, PRO.

51. Ibid., 16–18, 59–60; "Cable Crisis," 1490, Cable and Wireless Archives.

52. Walter Millis, *The Road to War: America, 1914–1917* (Boston: Houghton Mifflin, 1935), 147.

53. Haigh, *Cableships and Submarine Cables*, 330; Kunert, *Geschichte der deutschen Fernmeldekabel*, 349–360.

54. Ray S. Baker, *Woodrow Wilson and World Settlement* (Garden City, N.Y.: Doubleday, 1922), 2:468–469; John D. Tomlinson, *International Control of Radiocommunications* (Geneva: Journal de Genève, 1938), 47–48.

55. Committee of Imperial Defence, Imperial Communications Committee, "Memoranda," 1919, Cabinet Papers, CAB 35/2/87, PRO. The Admiralty's views were echoed in the instructions to the British delegation; see C.I.D., Imperial Communications Committee, "Proposed International Congress at Washington, Report of the Washington Congress Sub-Committee of Imperial Communications Committee (30 June 1920)," in Cabinet Papers, CAB 35/14, PRO.

56. Joseph S. Tulchin, *The Aftermath of War: World War I and U.S. Policy Toward Latin America* (New York: New York University Press, 1971), 224; Tribolet, *The International Aspects of Electrical Communications in the Pacific Area*, 234; Clark, *International Communications*, 165.

57. Frank J. Brown, *The Cable and Wireless Communications of the World: A Survey of Present Day Means of International Communication by Cable and Wireless, Containing Chapters on Cable and Wireless Finance* (London: Sir Isaac Pitman & Sons, 1927); Garratt, *One Hundred Years of Submarine Cables*, 36–52; Barty-King, *Girdle Round the Earth*, 180–202; Garnham

and Hadfield, *The Submarine Cable*, 161–167. See also Hugill, *Global Communications since 1844*, 48–51.

58. Committee of Imperial Defence, "Sub-Committee on Competition between 'Beam' Wireless and Cable Services (Brigadier General Sir S. H. Wilson, chair), Interim Report," Cabinet Papers, CAB 35/45, PRO; "Reports, Proceedings and Memoranda," Cabinet Papers, CAB 35/43, PRO.

59. "Strategical requirements in regard to cable communications," letter dated October 14, 1929, B1/112, Cable and Wireless Archives. Committee of Imperial Defence, Imperial Communications Committee, "Report on the Scheme Submitted by Imperial and International Communications, Limited, for the External Communications of the Colonies, Dependencies and Protectorates," July 28, 1930, Cabinet Papers, CAB 35/14, PRO.

60. Pascal Griset, "De la concurrence à la complémentarité: Câbles et radio dans les télécommunications internationales pendant l'entre deux guerres," *Recherches sur l'histoire des télécommunications* 2 (December 1988): 55–70.

61. Robert Sobel, *ITT: The Management of Opportunity* (New York: Times Books, 1982), 60–69.

62. Charles Graves, *The Thin Red Lines* (London: Standard Art Book Co., 1946), 23, 105–110, 127; Baglehole, *A Century of Service*, 23.

63. Graves, *The Thin Red Lines*, 68; Baglehole, *A Century of Service*, 24; Barty-King, *Girdle Round the Earth*, 280–285.

64. Ronald Lewin, *American Magic: Codes, Ciphers, and the Defeat of Japan* (New York: Farrar Straus Giroux, 1982), 88–109; David Kahn, *The Codebreakers: The Story of Secret Writing* (New York: Macmillan, 1967), 561–573; W. J. Holmes, *Double-Edged Secrets: U.S. Naval Intelligence Operations in the Pacific during World War II* (Annapolis, Md.: Naval Institute Press, 1979), 47, 82–97, 124.

65. Nigel West, *GCHQ: The Secret Wireless War 1900–86* (London: Weidenfeld and Nicolson, 1986), 155–156; Heinz Bonatz, *Seekrieg im Äther: Die Leistungen der Marine-Funkaufklärung 1939–45* (Herford: E. S. Mittler, 1981), 103–105.

66. Jaron Lanier, "Virtually There: Three-Dimensional Tele-Immersion May Eventually Bring the World to Your Desk," *Scientific American*, April 2001, 75.

10

Submarine Cables and Diplomatic Culture

David Paull Nickles

The Ever-Recurring Present

In the eyes of some, submarine telegraphy seems an esoteric subject, irrelevant to life in the twenty-first century. But consider this quotation: "The character of diplomacy . . . has undergone a great change. This change is threefold: first, as the result of the greater sense of the community of interests among nations . . . ; secondly, owing to the rise of democracy . . . ; thirdly, . . . due to modern means of communication." This statement could describe recent decades, in which "globalization" has increased interdependence among states, dictatorships have collapsed, and networked computers have facilitated the flow of information. In fact, it dates from the eleventh edition of *The Encyclopaedia Britannica*, published in 1910.[1] Our own experience of the present "communications revolution" is less different than we sometimes believe, and it shares much with earlier experiences of technological change.[2] When examining present international relations, we would do well to consider the earlier modifications in diplomacy associated with submarine telegraphy.

Telegraphy and Hopes of Peace

Those who saw the telegraph as an instrument of peace celebrated the expansion of submarine telegraphy. By improving communication among nations, cables, they believed, fostered mutual comprehension and respect and would ultimately eliminate the scourge of war.[3] John W. Brett, who participated in the laying of the first submarine cable across the English Channel, wrote, "Communication by submarine telegraph [between France and] England, although . . . still at the beginning of its development, already brings immense benefits to industry and amicable relations, which grow with the rapidity of the communications between the two countries."[4] Such views appeared even more often in the United States. After the laying of a transatlantic telegraph cable in 1858, a Boston clergyman noted that the effect of the telegraph is "to allay differences. Men who talk together daily cannot hate or disown one another."[5] Cyrus Field, an American businessman who played a leading part in laying that 1858 cable, later quoted the poet Whittier on the pacific tendencies produced by submarine telegraphy:

> Weave on, swift shuttle of the Lord,
> Beneath the deep so far,
> The bridal robe of earth's accord,
> The funeral shroud of war.[6]

A U.S. congressman confidently predicted that through the medium of interoceanic telegraphy "within a short period, the whole Christian world will be brought within the influence of a universal family circle, comprehending all forms of government, and all religions and creed, making up a vast community—one in everything that concerns political action, tariff regulations, coinage, currency, commerce, interchange of literature and the arts, material improvements, and in all the other agencies by which the world is moving onward to a higher destiny, and a more catholic spirit of fraternal accord."[7] These sorts of euphoric pronouncements are hardly unique. The notion that a new technology is remaking and improving the world will sound familiar to those who remember the frenzy surrounding the Internet during the 1990s. The tendency to direct hyperbolic praise toward technology has been common within the United States, where it draws strength from a tradition of religious millennialism and a robust belief in progress.[8]

Such hopes were not without foundation. Faster communication can sometimes further the cause of peace. When significant international events are occurring, timing can be crucial. Often, and especially when communication is relatively slow, the sequence in which historical actors learn of important events is more important than their actual chronology. For example, on Christmas Eve of 1814, British and U.S. negotiators concluded a treaty ending the war between their nations that had begun two and a half years earlier. News of the peace treaty did not, however, reach Washington until February of the next year. Meanwhile, on January 8, 1815, United States forces led by Andrew Jackson defeated British troops at the Battle of New Orleans. Faster communication (for example, a transatlantic telegraph cable) would have averted this engagement, saved many lives, and, in all probability, prevented Jackson from later becoming president of the United States.

In another instance, poorly timed news from a distant military theater undermined the prospect of peace negotiations. On September 24, 1865, Spanish naval forces off the coast of Chile announced that they would blockade that country's ports. As a result, Chile declared war on Spain. This war

quickly became unpopular in Spain, where the government's financial situation was calamitous. During November, British efforts to broker a peace showed promise. When Britain, with the support of France, made a mediation offer at the beginning of December, Spain quickly accepted. An end to the conflict seemed to be in sight. Unfortunately, the Chilean government and the Spanish naval forces blockading Chile had no way of knowing about the diplomatic progress occurring in London, which was connected to Madrid, but not to Santiago, by telegraph. Unable to communicate easily with his government, Spanish Admiral Pareja, the commander of his country's forces in the region, pursued high-handed and belligerent policies toward the Chileans and other South American governments. Near the end of November, two of his vessels blundered into Chilean traps and were captured. Humiliated, Pareja committed suicide. The peace process came unraveled with these events.[9] Spanish opinion was stunned and infuriated by the reverses suffered by their ostensibly superior forces. Their country's main motivations for fighting—thirsts for vengeance and lost prestige—intensified. Meanwhile, Chile, intoxicated by victory, felt less need than ever to make concessions to its enemy. Signs of Spanish weakness caused the war to expand further. In January 1866, Peru—followed soon after by Ecuador and Bolivia—entered the war on the side of Chile. This further complicated the diplomacy behind peace negotiations. Spain and Chile did not work out an armistice until years later, in April 1871. A peace treaty between the two countries was finally signed in June of 1883.[10] Earlier resolution of the conflict would have benefited both countries.

How did contemporaries expect international telegraphy to promote peace? In 1864 William M. Evarts, who became the U.S. secretary of state a decade later, predicted that the Atlantic cable would "be the great instrument in the removal of misunderstandings" between Britain and the United States.[11] Evarts's remarks deserve attention because of the influence that he wielded within the realm of international affairs. Moreover, he managed to articulate popular views about the telegraph more clearly than most of his contemporaries, thereby illuminating the underlying rationale. What was his basis for anticipating that the cable would produce an era of good relations between the two countries? First, he argued that telegraphy's ability to quickly correct misunderstandings would reduce conflict: "If a mistake should occur, we cannot long be deceived, nor does the false impression which it creates, become inveterate, and produce mischiefs which become irremediable before there is a possibility of explanation." Second, he asserted that the high cost of telegrams encouraged a plainspoken honesty: "We have a guaranty of truth in the necessary brevity of telegraphy, for superfluous falsehood is expensive."[12] Third, he argued that the ubiquity of telegraphic information would prevent privileged access to and monopolization of information on the part of those with a vested interest in fomenting conflict: "Whatever message or intelligence is transmitted, is made public in the same sense, in the same words, and in the same breath, all over the country, and cannot be made to assume the various forms which the letter writers of particular newspapers may, from interest or under dictation, put upon it."

The United States and Britain have not fought a war since the installation of the Atlantic cable, so Evart's prediction was indeed correct. Nonetheless, his reasoning is not convincing when examined more closely. Telegraphy had become widespread in both the United States and Europe by the 1850s and 1860s. Yet during this period the United States endured a civil war that resulted in the deaths of more than half a million people. James Gordon Bennett, the publisher of the *New York Herald*, declared in 1844 that the newly installed telegraph lines could "do more to guard against disunion . . . than all the most experienced, the most sagacious, and the most patriotic government, could accomplish."[13] This hope was not fulfilled. In the years that followed, greater "intensity of communication" between the North and South exacerbated, rather than ameliorated, strife between the two regions.[14] Telegraphy, by more tightly connecting distant sections of the United States, made manifest and worsened underlying tensions within American society and contributed to the political fragmentation of the 1850s.[15]

Likewise, telegraphy did not pacify Europe, which experienced a string of wars (five between 1854 and 1870) after having had many years of peace during the era before the establishment of electric telegraphy. The 1870 Franco-Prussian War, whose immediate origins are associated with the famous "Ems telegram," strikingly contradicts Evarts's optimistic prognostications of a few years earlier. Prussian leader Otto von Bismarck had received a telegram describing a meeting at which the King of Prussia had politely refused demands from the French ambassador when the two men met at the spa of Ems. Bismarck edited out aspects of politeness in the king's refusal, making it seem much more brusque and insulting than it was. He then transmitted the new version of the encounter to the press. The resulting furor led the French government to declare war against Prussia.[16]

This example demonstrates that telegraphy could rapidly propagate a false version of events. Indeed, the misleading account that Bismarck released to the press traveled more quickly than the fuller and more accurate version that the French ambassador sent to Paris through diplomatic channels. Moreover, Bismarck's sensationalistic story of an insulting encounter served the interest of the press better than the true, and more innocuous, account.

Furthermore, Bismarck's condensed version was less accurate than the longer account that he had received. "The necessary brevity of telegraphy" turned out to be no "guaranty of truth." Finally, Bismarck, from his position of power, proved himself quite successful at manipulating information flows in order to foster antagonism between France and Prussia.[17] Telegraphy did not provide everyone with an equal opportunity to influence the exchange of news.

Diplomatic Suspicion of the Telegraph

Foreign policy institutions did not rush to embrace telegraphy. One historian has declared of the early-twentieth-century French foreign ministry, "In its daily routine the Quai d'Orsay frequently seemed to regard the typewriter as a new-fangled instrument and the telephone and telegraph as gadgets whose usefulness had not as yet been fully demonstrated."[18] What accounted for the conservatism and suspicion of innovation found in most late-nineteenth-century foreign ministries? One could dismiss such tendencies as merely the inertia that encumbers many bureaucracies. But there were additional reasons for diplomats to view telegraphic communication with skepticism. First, the telegraph seemed to threaten their ability to perform their jobs effectively; second, it undermined their lifestyle; and third, it reduced their freedom.

The first of these reasons suggests that telegraphy impaired the ability of diplomats to resolve unnecessary conflicts between states. Many diplomats believed that the passage of time was their ally and that a technology that accelerated events presented more of a danger than an opportunity. In the era before the electric telegraph, Charles-Maurice de Talleyrand-Périgord, a brilliant French diplomat and statesman, told subordinates to wait two days before sending off his dispatches so that he might have time to change his mind.[19] When Bernhard von Bülow began his diplomatic career at the German foreign office, his father, who was foreign secretary, advised him, "Above all, don't let yourself be hurried. . . . Don't overdo it [*Pas trop de zèle*] is a golden rule when rightly understood."[20] Other contemporary observers agreed that speed was often harmful in diplomacy. At the time of the *Alabama* negotiations (involving the United States and Britain) during the early 1870s, the *Economist* asserted, "We are over-working the powers of the telegraph when we try to do by its agency those delicate and complicated pieces of business which require either full and free personal intercourse, or very deliberate judgment, or both. There are many matters in which time is of far less importance than care, and in which the natural impatience of human nature in general, and of political nature in particular, ought to be steadily resisted and not yielded to."[21]

Diplomats contended that the performance of their duties depended on frequent opportunities for relaxed fraternizing. Bustling about at the beck and call of the telegraph endangered their effectiveness. In order to interact meaningfully with European elites, diplomats needed to behave as the elites did and devote much of their time to the pursuit of pleasure. This was an expensive proposition. In February 1853 the British ambassador to France insisted that lavish spending on "entertainment" was required to "sustain with credit that position in society which is inseparably connected with the rank of Ambassador + which is almost as essential to him as a good understanding with the govt. to which he is accredited."[22] Much of what looked like play was actually work—or so diplomats claimed. A German envoy recalled, "As part of our duties, we were expected to go out [and participate in society] as much as possible."[23] One of his American peers claimed that his years of fox-hunting "were the best preparation which I could have had" for a diplomatic career in England.[24] In part, social activities provided a good opportunity for collecting important information. In part, they allowed diplomats to make friends for their country and promote amicable relations between nations. For example, Herbert Bismarck, the son of the German chancellor, captivated Russian society with his enormous alcohol tolerance.[25]

Questions of functionality aside, telegraphy threatened the relaxed and gentlemanly lifestyle that diplomats preferred.[26] Foreign ministries across Europe drew heavily from the nobility when filling positions.[27] With the decline of aristocratic dominance in government, diplomacy became the last refuge of an "aristocrat who knew his times."[28] The aristocracy tended to distinguish itself from the nouveau riche by emphasizing its disdain for paid labor. It is, therefore, hardly surprising that, as one Austro-Hungarian envoy said about his peers, few diplomats displayed much "discipline and enthusiasm for work."[29] Reflecting on such comments, a historian of the Habsburg diplomatic service writes, "Those diplomats who showed themselves too industrious or failed to cover for their dilatory comrades made themselves unpopular."[30]

Unlike its European counterparts, the U.S. State Department lacked an authentic aristocracy from which to recruit. It nonetheless managed to choose diplomats with characteristics rather similar to those of their European peers.[31] Although the State Department at times tried to encourage industriousness on the part of its employees, it did not meet with great success during the nineteenth and early twentieth centuries.[32]

The importance of dance in diplomatic life is indicative of the priority placed on leisure activities. During an era when dance steps tended to be relatively complex and regimented, skill on the parquet was generally a sign of upper-class status. Leisured individuals were much more likely to possess the abundant time and money required to master elaborate dance etiquette.[33] Diplomats argued that attendance at balls was an essential aspect of their duties. Bernhard von Bülow, on joining the German diplomatic service, was advised by a more senior diplomat that, to achieve success, he should "go to parties as much as possible and dance a great deal, as this would make [him] popular."[34] When Count Heinrich Lützow, an Austro-Hungarian envoy, first arrived at his country's foreign ministry to begin his career, Baron Leopold von Hofmann, one of the highest-ranking officials, asked him, "Can you lead off at a ball?" Receiving an affirmative response, Hofmann exclaimed, "Perfect, because that's the main thing!"[35] Some years later, when Lützow had risen to the rank of ambassador to Italy, he remonstrated against an effort to transfer Baron Georg Franckenstein, his best leading dancer (*Vortänzer*), before the end of carnival. When Franckenstein did depart, Lützow requested that another excellent dancer replace him.[36] Examples abound of the emphasis on leisure in pre–World War I diplomacy. Diplomats sought to preserve this genteel and hedonistic lifestyle in the face of onrushing demands, aptly symbolized by telegraphy.[37]

Telegraphy also threatened the independence of diplomats. System builders of all sorts seized on the telegraph as a tool for extending their authority. In fields such as imperialism and war, governments adopted new forms of communication in an effort to strengthen their control. Lord Curzon, the viceroy of British India, toiled to reduce "cable rates to Europe," on the theory that "a cheap tariff is the greatest instrument of Imperial unification that can be devised."[38] At the same time, the cable connection to London caused Curzon frustration over his own loss of independence. Other local colonial officials shared his vexation with telegraphic oversight.[39] The same phenomenon occurred as distant military theaters came under the immediate command of civilians. One of the earliest witnesses to this development was General Simpson, the British commander in chief during the Crimean War, who reportedly remarked, "The confounded telegraph has ruined everything."[40]

In the same manner, telegraphy reduced the autonomy and power of diplomats. Canadian Prime Minister Pierre Trudeau once declared that the foreign service possesses "diminished relevance in an era of instantaneous world-wide communications, in which there is increasing reliance on personal contacts between senior members of governments."[41] Many years earlier, Lord Salisbury, one of Britain's most effective foreign ministers of the nineteenth century, exploited submarine telegraphy to keep tabs on distant diplomats. When plagued by unruly consuls in Samoa, he advocated building a telegraph line to this trouble spot in order to monitor their activities.[42] In the case of T. G. Knox, Britain's diplomatic representative to Siam from 1878 to 1879, there was already a cable in place. At this time, Knox's willful daughter, "Miss Fanny," eloped with a Siamese nobleman, becoming one of his many wives. Believing that his daughter had compromised herself, Knox gave his consent to the union. The Siamese government, however, took exception to the nobleman's haughty treatment of a foreign dignitary's daughter, especially since the nobleman was also accused of having embezzled £200,000 from gold mines that he oversaw. As punishment, the nobleman was condemned to receive thirty stripes in a public flogging. Rather than seeing this sentence as a vindication of his honor, Knox demanded that the Siamese government apologize and immediately liberate his new son-in-law. When the Siamese government balked at this directive, Knox called up a British gunboat. Salisbury learned of the controversy and, interceding by telegraph, recalled Knox before he could take rash action. Submarine telegraphy made possible prompt action to restrain an impetuous diplomat on the other side of the globe.[43]

Like Lord Curzon and General Simpson, some diplomats felt emasculated by the centralizing tendencies of the telegraph. Ernest Satow, a British envoy, noted, "In former times a wide discretion in the interpretation of his instructions was permitted to an envoy, in case it became necessary to take a sudden decision." But, in the age of the telegraph, Satow reported, diplomats could contact their superiors if their instructions seemed inappropriate to the situation at hand. They lost much of their freedom to take independent action based on their own judgment.[44] Many envoys worried that telegraphy trivialized their role in the execution of foreign policy. Arthur Ponsonby, a disaffected young British diplomat, complained, "The surprising progress made in the means of communication has deprived diplomats of many of the grave responsibilities with which they were formerly charged."[45] An American consul declared, "The office of ambassador or minister is in most cases not so important as it once was; in these days of rapid communication a large part of the relations of two governments may be conducted practically direct between the two foreign offices."[46] Diplomats jealous of their independence, their lifestyle, and their timeworn techniques found much to dislike in diplomacy by telegraph.

Issues of Cost and Reliability

Fortunately for diplomats, they had some ready excuses that allowed them to avoid use of the telegraph. The high cost of telegraphy and its unreliability provided them with justifications for not employing it.[47] While these two problems occurred with messages conveyed by landlines, they were particularly pronounced with cablegrams, which were much costlier[48] and generally more prone to garbling (due in part to seawater's ability to conduct electricity, which smeared signals in an effect called capacitance). Both of these phenomena had considerable influence on the form in which telegrams were actually received.

First, the issue of cost. The physicist William Thomson received commendation on literary grounds for his research, which greatly advanced the development of submarine telegraphy. As one tribute noted, Thomson's cable was so exorbitant that it encouraged succinctness on the part of all those who used it, enforcing the precept "brevity is the soul of wit."[49] Diplomats too were affected by telegraphy's tendency to inspire concision. The British ambassador to France remarked, "These dispatches being sent by telegraph were naturally short."[50] Of particular interest is the extent to which telegraphy tended to homogenize style across foreign ministries.[51] For example, in general British dispatches were known for their pithiness, while French dispatches were renowned for their fluency.[52] Did telegraphy blur such differences? More research needs to be done in order to test such hypotheses. Yet, preliminary evidence suggests that although telegraphy created similarities in diplomatic dispatches across foreign ministries, some, especially France's Quai d'Orsay, were more resistant than others to the emergence of the "telegraphic style."[53]

As years passed, foreign ministry budgets grew, the carrying capacity of telegraph lines increased, and the per-word cost of telegrams diminished.[54] Diplomats became more accustomed to using the telegraph. Yet, although old restrictions about the use of the telegraph loosened, they did not disappear. Lengthy, ruminative telegrams still gained notice, frequently provoking irritation in those who handled and read them. But on occasion, when they had something important to say, diplomats could turn this attention-getting tactic to their advantage. At the end of World War II, George Kennan, in charge of the American embassy in Moscow, found himself frustrated by his inability to influence U.S. policy makers back in Washington. He felt as though he were "talking to a stone."[55] He broke out of this impasse by violating the rules about diplomatic communication via the telegraph. On February 22, 1946, he sent a telegram of such unusual length (5,540 words) that it has since become immortalized as the "long telegram."[56] At the time, Kennan apologized "for this burdening of [the] telegraphic channel," but explained that the "questions involved are of . . . urgent importance" and "I cannot compress answers into [a] single brief message without yielding to what I feel would be [a] dangerous degree of oversimplification."[57] Kennan's sense of mission did not endear him to others at the embassy, particularly Martha Mautner, the embassy employee who received the thankless task of encoding the massive communication. She had a "heavy date" that evening and planned to attend a dance with her companion at one of the other embassies. Without success, she tried to convince her prickly boss that no harm would come from a day's delay.[58] Kennan's stratagem worked. The length of his cablegram captured attention; the appeal of his message, which predicted the inevitability of Soviet hostility toward the United States for the foreseeable future, allowed him to avoid reprimand and, in fact, won him praise.[59] Historians credit (or blame, as the case may be) the "long telegram" with having set the course of U.S. foreign policy toward the Soviet Union during the Cold War period.[60]

Second, the issue of reliability.[61] Telegraphy encountered an early trial by fire during the Crimean War. The British and French governments laid one of the early submarine telegraph cables in an effort to improve communication with their theater military commanders. By April 1855 it was possible to send a telegram to the Crimea from Paris or London.[62] Perhaps for the first time, governments sought to micromanage a war from many hundreds of miles away. The results were less than felicitous. Soldiers found themselves baffled by the cablegrams they received. On July 17, 1855, the British commander in chief in the Crimea, General Simpson, complained to his superior in London, "I think, my Lord, that some telegraphic messages reach us that cannot be sent under due authority and are perhaps unknown to you, although under protection of your Lordship's name." By way of example, Simpson described being awoken in the middle of the night by a dragoon bearing an urgent question: "Lord Panmure to General Simpson—Captain Jarvis has been bitten by a centipede. How is he now?" Simpson believed that this cablegram addressed "rather too trifling an affair to call for a dragoon to ride a couple of miles in the dark, that he may knock up the Commander of the Army out of the very small allowance of sleep permitted him!" But what really irked Simpson was the discovery, "upon sending in the morning another mounted dragoon to inquire after Captain Jarvis, four miles off, . . . that he never has been bitten at all, but has had a boil, from which he is fast recovering."[63]

A more dramatic deception occurred in the Australian town of Adelaide, which had received a connection to the international telegraph grid in 1872. Among the first telegrams it received from London was a fallacious report that war would soon break out between Britain and the United States. In the midst of this suspense, the wire failed and was not again in working order for several weeks.[64] Such occurrences were maddening. Some years earlier, in 1861, a British diplomat had complained before a parliamentary committee that telegraphy was a medium "that first appears to give information but which often turns out to be totally at variance with the facts as they come out afterwards when you receive the post."[65] Submarine telegraphy showed a conspicuous facility for conveying misinformation and confusion over vast distances with unparalleled speed.[66]

Conclusion

Ultimately, the exploration of history is valuable at least in part because it makes human beings more sensitive to the range of choices that they confront. For example, it can broaden our frame of reference and allow us to think outside the box of established patterns of thought concerning international problems. Many well-placed observers have argued that technological change threatens to overwhelm our institutions for dealing with diplomatic crises. Then-Secretary of State Henry Kissinger expressed our predicament in an address to the General Assembly of the United Nations: "We are, in fact, members of a community drawn by modern science, technology, and new forms of communication into a proximity for which we are still politically unprepared. Technology daily outstrips the ability of our institutions to cope with its fruits."[67] We might consider the example of the founders of the League of Nations. During the Great War, many observers believed that in July 1914 the pace of events—which moved at the speed of the railroad and telegraph—had overwhelmed the ability of diplomatic institutions to resolve disputes. The League of Nations was created in part as a response to this breakdown. At the League, preassembled delegates sat in continual readiness to address incipient crises.[68] The League ultimately proved to be an abysmal failure, although this was perhaps more a result of the foolish and destructive behavior of the world's great powers than of insuperable flaws within the institution itself. But though their efforts failed, we can still take inspiration from the creativity and energy of the statesmen involved in this effort. As Kissinger declared, "Our political imagination must catch up with our scientific vision."

Notes

The views expressed in this essay are those of the author and are not necessarily those of the Department of State or the U.S. Government. I would like to thank Adam Albion, Becky Carpenter, John Lewis Gaddis, David Geyer, Tom Tulenko, and the participants of the Dibner/Smithsonian conference on submarine cables.

1. *Encyclopaedia Britannica*, 11th ed., s.v. "Diplomacy."

2. For historical perspective on the notion of a "communications revolution," see Richard R. John, "American Historians and the Concept of the Communications Revolution," in *Information Acumen: The Understanding and Use of Knowledge in Modern Business*, ed. Lisa Bud-Frierman (London: Routledge, 1994).

3. Geoffrey Blainey, *The Causes of War* (London: Macmillan, 1988), 18–23.

4. John W. Brett to French Minister [of the Interior?], March 18, 1853, Affaires diverses politiques, Angleterre, 18, Archives of the French Foreign Ministry, Paris.

5. Ezra Stiles Gannett, *The Atlantic Telegraph: A Discourse Delivered in the First Church, August 8, 1858* (Boston: Crosby, Nichols, 1858), 14. See also William Buell Sprague, *A Sermon Addressed to the Second Presbyterian Congregation, Albany, on Sunday Morning, September 5, 1858 on the Completion of the Atlantic Telegraph* (Albany: Charles van Benthuysen, 1858), 24–25.

6. *Twenty-Seventh Anniversary of the First Atlantic Cable: Mr. Cyrus W. Field's Banquet at the Star and Garter Hotel, Richmond, on August 5th, 1885* (London: Gilbert & Rivington, 1886). Collection 205, Western Union Telegraph Company, box 142, 1993 Addendum, Cyrus Field Papers, National Museum of American History Archives Center, Smithsonian Institution, Washington, D.C.

7. "Telegraph with Foreign Countries," H.Rp. 35, Report submitted by Representative Fernando Wood in the House Committee on Foreign Affairs to accompany bill H.R. 1263, 41st Cong., 2d sess., 1870, 1–2.

8. Daniel J. Czitrom, *Media and the American Mind: From Morse to McLuhan* (Chapel Hill: University of North Carolina Press, 1982), 9–10. Note also Leo Marx's concept of a "technological sublime" that has often found expression in American culture. Leo Marx, *The Machine in the Garden: Technology and the Pastoral Ideal in America* (New York: Oxford University Press, 1970), 194, 230.

9. Stanley to Crampton, copy, August 15, 1866, Foreign Office Files, FO 115/451, Public Record Office, London (hereafter PRO).

10. Joseph Smith, *Illusions of Conflict: Anglo-American Diplomacy toward Latin America, 1865–1890* (Pittsburgh: University of Pittsburgh Press, 1979), 50–57; William Columbus Davis, *The Last Conquistadores: The Spanish Intervention in Peru and Chile, 1863–1866* (Athens: University of Georgia Press, 1950); James W. Cortada, "Diplomatic Rivalry between Spain and the United States over Chile and Peru, 1864–1871," *Inter-American Economic Affairs* 27 (Spring 1974): 47–57; Simon Collier and William E. Sater, *A History of Chile, 1808–1994* (Cambridge: Cambridge University Press, 1996), 117–118.

11. *Banquet Speech from Friday, 15 April 1864, Europe and America: Report on the Proceeding at an Inauguration Banquet Given by Mr. Cyrus Field* (London: Wm. Brown, 1864), Collection 205, box 142, 1993 Addendum, Cyrus Field Papers, National Museum of American History Archives Center. All of the quotes from Evarts that follow are from this pamphlet.

12. For a similar argument from an American clergyman, see Gannett, *The Atlantic Telegraph*, 12.

13. Thomas R. Hietala, *Manifest Design: Anxious Aggrandizement in Late Jacksonian America* (Ithaca: Cornell University Press, 1985), 196.

14. David Potter, "Historians and the Problem of Large-Scale Community Formation," in *History and American Society*, ed. Don E. Fehrenbacher (New York: Oxford University Press, 1973), 58–59.

15. Menahem Blondheim, *News over the Wires: The Telegraph and the Flow of Public Information in America, 1844–1897* (Cambridge, Mass.: Harvard University Press, 1994), 191–194. The theory that easy intercourse promotes harmony has been applied to other events in American history. For a critique of the claim that poor transatlantic communication resulted in the American Revolution and the breakup of the first British Empire, see Ian K. Steele, *The English Atlantic, 1675–1740: An Exploration of Communication and Community* (New York: Oxford University Press, 1986), 273; and I. K. Steele, "Moat Theories and the English Atlantic, 1675–1740," in the *Canadian Historical Association's Historical Papers, 1978* (1979): 18–33.

16. Lothar Gall, *Bismarck: Der weiße Revolutionär* (Frankfurt: Ullstein, 1980), 434; Pierre Renouvin, *Le XIXe siècle*, I., *De 1815 à 1871: L'Europe des nationalités et l'éveil de nouveaux mondes*, Histoire des relations internationales (Paris: Hachette, 1994), 2:631.

17. This sort of behavior does not appear to have been exceptional for Bismarck. He has a reputation for having planted fabricated articles in the press. Ernest R. May, "The News Media and Diplomacy," in *The Diplomats, 1939–1979*, ed. Gordon A. Craig and Francis L. Loewenheim (Princeton: Princeton University Press, 1994), 667.

18. Richard D. Challener, "The French Foreign Office: The Era of Philippe Berthelot," in *The Diplomats, 1919–1939*, ed. Gordon A. Craig and Felix Gilbert (New York: Atheneum, 1963), 1:60.

19. Preface by Jacques Chastenet, in *Pierre de Margerie (1861–1942) et la vie diplomatique de son temps*, ed. Bernard Auffray (Paris: C. Klincksieck, 1976), 9.

20. Bernhard von Bülow, *Memoirs of Prince von Bülow* (Boston: Little, Brown, 1932), 4:300. The expression "pas trop de zèle," one of the most famous maxims in diplomacy, is associated with Talleyrand. See also Lord Strang et al., *The Foreign Office* (London: George Allen & Unwin, 1957), 98.

21. Quoted by Robert J. Cain, "Telegraph Cables in the British Empire, 1850–1900" (PhD diss., Duke University, 1971), 187–188.

22. Cowley to FO [John Ampthill], "Separate," February 7, 1853, Foreign Office Files, FO 519/1, PRO.

23. Richard von Kühlmann, *Erinnerungen* (Heidelberg: Lambert Schneider, 1948), 122.

24. Allan Nevins, *Henry White: Thirty Years of American Diplomacy* (New York: Harper, 1930), 28–29.

25. Kühlmann, *Erinnerungen*, 145.

26. Research on Britain's Colonial Office suggests that telegraph cables produced a busier, less carefree, more bureaucratic office culture. Brian Layton Blakeley, "The Colonial Office: 1870–1890" (PhD diss., Duke University, 1966), 102, 301–302.

27. On the German foreign ministry, see Wilhelm von Schoen, *The Memoirs of an Ambassador: A Contribution to the Political History of Modern Times*, trans. Constance Vesey (London: George Allen & Unwin, 1922), 121.

28. R. J. B. Bosworth, *Italy, the Least of the Great Powers: Italian Foreign Policy before the First World War* (London: Cambridge University Press, 1979), 94.

29. William D. Godsey Jr., *Aristocratic Redoubt: The Austro-Hungarian Foreign Office on the Eve of the First World War* (West Lafayette, Ind.: Purdue University Press, 1999), 201. See also Bülow, *Memoirs of Prince von Bülow*, 325; Emerich Csáky, ed., *Vom Geachteten zum Geächteten: Erinnerungen des k. und k. Diplomaten und k. ungarischen Außenministers* (Vienna: Böhlau, 1992), 129; Heinrich Kanner, *Kaiserliche Katastrophen-Politik: Ein Stück zeitgenössischer Geschichte* (Leipzig: E. P. Tal, 1922), 87–88.

30. Godsey, *Aristocratic Redoubt*, 200.

31. Richard Hume Werking, *The Master Architects: Building the United States Foreign Service, 1890–1913* (Lexington: University of Kentucky, 1977), 121; Henry Kissinger, *Years of Upheaval* (Boston: Little, Brown, 1982), 443.

32. For example, Secretary of State Hamilton Fish attempted in 1869 to establish regular working hours at the State Department (Fish circular of October 20, 1869, vol. 1, E726, RG 59, National Archives and Records Administration, College Park, Md.). In 1910 the State Department strove to gain greater control over its diplomats by issuing two memoranda about their obligation to inform the department of their whereabouts (ibid., vol. 7).

33. On the use of etiquette as a marker of class status, see Thorstein Veblen, *The Theory of the Leisure Class* (1899; repr., New York: Penguin, 1994), 48–49. Prince Bülow noted that during the Berlin winter season of 1873–1874, popular dances included the quadrille, the lancers, the gallop, the polka, and the waltz. "There were no atrocities like the two-step, the fox-trot and shimmy." Bülow, *Memoirs of Prince von Bülow*, 314.

34. Bülow, *Memoirs of Prince von Bülow*, 298.

35. Godsey, *Aristocratic Redoubt*, 26.

36. Ibid., 26–27.

37. For a literary depiction of the perceived association between telegraphy and the industrial bourgeoisie, see E. M. Forster's novel *Howard's End*, which identified the sending of telegrams as a trait of the grasping, mercantile, and philistine Wilcox family. Forster, *Howard's End* (1910; New York: Bantam Books, 1985), 20, 81. For more on the tension between technology and pre-1914 diplomatic culture, see David Paull Nickles, *Under the*

Wire: How the Telegraph Changed Diplomacy (Cambridge, Mass.: Harvard University Press, 2003), chap. 5.

38. Lord Curzon's farewell address, Bombay Chamber of Commerce, November 8, 1905, in *Lord Curzon in India*, ed. Thomas Raleigh (London: Macmillan, 1906), 285. See also James W. Carey, *Communication as Culture: Essays on Media and Society* (Boston: Unwin Hyman, 1989), 212.

39. Paul M. Kennedy, "Imperial Cable Communications and Strategy, 1870–1914," *English Historical Review* 86 (1971): 751.

40. Jeffrey Kieve, *The Electric Telegraph: A Social and Economic History* (Newton Abbot: David & Charles, 1974), 240.

41. Hector Mackenzie, "Different Country, Same Problems," *Foreign Service Journal* 76, no. 12 (December 1999): 32.

42. George Herbert Ryden, *The Foreign Policy of the United States in Relation to Samoa* (New Haven: Yale University Press, 1933), 447–448; Salisbury to Malet, April 24, 1889, in *Lady Gwendolen Cecil's Life of Robert, Marquis of Salisbury* (London: Hodder and Stoughton, 1932), 4:128. Salisbury may also have been seeking to extend the British cable network.

43. Raymond A. Jones, *The British Diplomatic Service, 1815–1914* (Waterloo, Canada: Wilfred Laurier University Press, 1983), 213.

44. Ernest Satow, *A Guide to Diplomatic Practice* (London: Longmans, Green, 1917), 1:143.

45. Arthur Ponsonby to Thomas Sanderson, "Arthur Ponsonby, the Diplomatic Service and the Foreign Office, 1900–1902," quoted in Zara Steiner, *The Foreign Office and Foreign Policy* (London: Ashfield, 1986), appendix 4, 227.

46. William Fenwick Harris, "The Consular Service as a Profession," *Harvard Advocate* (c. 1911), 67–71, available at Harvard University, Widener Library, under OW:Int 6080.1.

47. For an exploration of how diplomats circumvented use of the telegraph, see Nickles, *Under the Wire*, chap. 7.

48. The telegraph rates for communicating with American embassies in 1912 are available at 119.21/73, 1910–29, RG 59, National Archives and Records Administration.

49. Toast by Sir William Stirling-Maxwell, in Silvanus P. Thompson, *The Life of William Thomson, Baron Kelvin of Largs* (London: Macmillan, 1910), 1:554.

50. Cowley to Clarendon, #397, June 6, 1853, Foreign Office Files, FO 519/2, PRO.

51. The extent to which telegraphy standardized language across foreign ministries is related to broader trends, which operated on entire nations around the globe. Among contemporaries, one heard the prediction that telegraphy was a tool of God transforming the world into "one nation, speaking one language." Annteresa Lubrano, *The Telegraph: How Technology Innovation Caused Social Change* (New York: Garland, 1997), 126. Clearly, this has not happened, although electric media are contributing to the destruction of the vast majority of the world's languages.

52. Mountague Bernard, *Four Lectures on Subjects Connected with Diplomacy* (London: Macmillan, 1868), 123.

53. Jacques Baeyens, *Au bout du quai: Souvenirs d'un retraité des postes* (Paris: Fayard, 1975), 217. My own study of French diplomatic telegrams, impressionistic though it is, leads me to believe that they were more elegantly written and less likely to drop articles than were their American counterparts. Illustrative examples are available at Correspondance Politique et Commerciale, 1897–1918, Nouvelle série, Grande-Bretagne, vols. 21 and 22, Archive of the Ministère des Affaires Étrangères, Paris.

54. On the decline of prices for cablegrams, see Jorma Ahvenainen's essay in this volume.

55. George F. Kennan, *Memoirs, 1925–1950* (Boston: Little, Brown, 1967), 293.

56. Walter Isaacson and Evan Thomas, *The Wise Men: Six Friends and the World They Made* (New York: Simon and Schuster, 1986), 352.

57. George Kennan, "The Kennan 'Long Telegram,'" in *Origins of the Cold War*, ed. Kenneth M. Jensen (Washington, D.C.: USIP Press, 1993), 17.

58. Interview with Martha Mautner, in "Iron Curtain, 1945–1947," vol. 1, no. 2 of *Cold War*, a Jeremy Isaacs production for Turner Original Productions, 1998.

59. Kennan, "The Kennan 'Long Telegram,'" 295.

60. John Lewis Gaddis, *The Long Peace: Inquiries into the History of the Cold War* (New York: Oxford University Press, 1987), 39; Melvyn P. Leffler, "Commentary," in Jensen, *Origins of the Cold War*, 81.

61. On the unreliability of submarine cables within the context of British colonial policy, see Blakeley, "The Colonial Office: 1870–1890," 110n3.

62. Daniel R. Headrick, *The Invisible Weapon: Telecommunications and International Politics, 1851–1945* (New York: Oxford University Press, 1991), 17.

63. General Simpson to Lord Panmure, Crimea, "Private," July 17, 1855, in *The Panmure Papers*, ed. George Douglas and George Dalhousie (London: Hodder and Stoughton, 1908), 1:294–296; Lytton Strachey, *Eminent Victorians* (New York: G. P. Putnam's Sons, n.d. [1922?]), 168. My thanks to Adam Albion for bringing this story to my attention.

64. Headrick, *The Invisible Weapon*, 40.

65. Jones, *The British Diplomatic Service*, 124.

66. On this point, see John A. Britton, "'The Confusion Provoked by Instantaneous Discussion': The New International Communications Network and the Chilean Crisis of 1891–1892 in the United States," *Technology and Culture* 48 (2007): 756.

67. Kissinger, *Years of Upheaval*, 447. Whatever one thinks of the policies that Kissinger pursued while in office, he was an intelligent and highly placed observer of international affairs. For similar comments from well-known diplomats, see U. Alexis Johnson, *The*

Right Hand of Power (Englewood Cliffs, N.J.: Prentice-Hall, 1984), 633–634; and Strang, *The Foreign Office*, 197.

68. Henry Kissinger, *Diplomacy* (New York: Simon and Schuster, 1994), 213, 215, 217; Strang, *The Foreign Office*, 40; John I. Knudson, *A History of the League of Nations* (Atlanta: Turner E. Smith, 1938), 137; A. J. P. Taylor interview in Norman F. Cantor's *Perspectives on the European Past: Conversations with Historians*, part 2 (New York: Macmillan, 1971), 274; Keith Hamilton and Richard Langhorne, *The Practice of Diplomacy: Its Evolution, Theory and Administration* (London: Routledge, 1995), 160.

11

Submarine Cables and the Two Japanese Empires

Daqing Yang

> It is no exaggeration that the development of submarine cables—once the dominant means of international communication—followed the same trajectory as Japan's modern history in international affairs and shared the ebb and flow of our national fortune. To follow the history of the submarine cable enterprise, therefore, is to narrate the political and economic development of our country at the same time.
>
> Ôya Harukazu, Director, NTTPC Submarine Cable Equipment Co.[1]

Harold A. Innis, the late Canadian economic historian and possibly the greatest theoretician of communication studies, was among the first to point out perhaps the obvious: the geographical limits of empires were determined by the possibilities for effective communication. Consequently, changes in the technology of transport and communication in history permitted vast changes in the possibilities for the extension of empires. Surprisingly, however, the link between modern imperialism and communication has been largely overlooked by historians until recently. Whereas telegrams and other official correspondence have always been an invaluable source to students of international history, by and large the communication process as well as the telecommunication network itself has often been taken for granted except in such matters as code breaking. In *Communication as Culture*, James Carey complained, "While the growth of empire and imperialism have been explained by virtually every possible factor, little attention has been paid to telegraphy in generating the ground conditions for the urban imperialism later in the century."[2]

This has begun to change. Given the growing realization of the importance of communications technology, recent years have seen a number of histories describing connections between communications technology and imperialism. The electric telegraph, along with the invention of the telephone in the 1870s, was enthusiastically welcomed by empire builders. Just as couriers riding on horseback served the communication needs of empires for centuries, telegraph cables, as historian Daniel Headrick has pointed out, now formed "an essential part of the new imperialism."[3] Headrick considers telecommunication among the several key technological developments that played the most crucial role in European imperialism, "either by making imperialism possible where it was otherwise unlikely, or by making it suitably cost-effective in the eyes of budget-minded governments." Telecommunications, argues Headrick, gave value to a handful of mostly deserted islands in the most isolated parts of the world and, in a few instances, helped empires to expand. But more important was the fact that cables and wireless tied the European empires together. In times of peace they were the lifelines of the ever-increasing business communication that bound imperialist nations to their colonies around the world. In times of crisis, they were valuable tools of diplomacy. And in times of war, they were security itself. He further distinguishes several stages in European imperialism that began with penetration and were followed by conquest. In the final phase of consolidation, he notes, "the links that tied the colonies to Europe and promoted their economic exploitation included steamship lines, the Suez Canal, the submarine telegraph cables, and the colonial railroads."[4] As Headrick phrases it, "The web of power that tied the colonial empires together was made of electricity as well as steam and iron."[5]

In most recent works on the role of telecommunications in world politics, however, Japan—the only non-Western power to have built a modern empire—is conspicuously absent. This chapter seeks to fill a small gap by looking at two distinct phases of Japanese empire building in which submarine cables played prominent roles. In doing so, it examines a number of empirical questions: How did Japanese policy makers come to view telecommunications as an essential component in Japan's overall strategy? How did they conceive of a Japanese hegemony in telecommunications in East Asia, and where did they find inspiration? How did such a telecommunications network fit into Japan's vision of empire? What was distinctive about this network? What resources—technological, financial, and human—enabled Japan to build a telecommunications network of such a scale? How successful was Japan in achieving its goals?

Telegraph Comes to Japan

After the Tokugawa shogunate in Japan in the 1850s gave in to Western pressure to set up diplomatic representation and settlement in its ports, constant international communication became a necessity. As part of the unequal treaties, the United States, Britain, and France insisted on maintaining their own postal services in the settlements in Japan, as they had done in China earlier, on the grounds that the Japanese service was of an unacceptable standard. Western shipping companies, through their government-subsidized steamship routes, linked the once remote island country to a global information system.

In the meantime, the new form of rapid communication was extending its tentacles and preparing to envelop the Far East. In the decades following Samuel Morse's 1844 experiment between Washington, D.C., and Baltimore, the world was experiencing what could be only described a global communications revolution. In 1851 the first operational submarine cable was laid across the English Channel between Calais and Dover. By the end of the decade the trunk line network in Europe was complete. The 1860s saw rapid expansion, though often with failure, of telegraphic lines outside Europe. In 1864 the Indo-European Telegraph Company laid a cable in the Persian Gulf, thereby connecting—partly over land and partly under water—Britain and the Crown colony of India. Six years later, in 1870, the British Indian Company connected Alexandria and Bombay. This line made possible an all-cable route. Competition was rampant not only between actual lines in operation but also between attempts to gain concessions from various foreign governments to land submarine cables or to build landlines. As a result, pooling and merger became common. In 1869 the Great Northern Telegraph Company was founded following the merger of three companies that engaged in telegraphic communication in northern Europe. Registered in Copenhagen, Denmark, it had a capital of £400,000, part of which was held in London. In 1873 three British companies operating in Asia merged to found the Eastern Extension, Australasia and China Telegraph Co. Ltd. Immediately seeking to extend telegraph communications to China and Japan by way of Siberia and India, respectively, these two telegraph companies were to dominate international communication in East Asia for decades to come.[6] From the very beginning, international telecommunication was as much to inform as to control, and the burgeoning network thus had a quality of unevenness characteristic of European expansion.

In September 1870, the three-year-old Meiji government in Japan entered into an agreement with a Danish diplomat representing the Great Northern Telegraph Company, by which the company was to construct submarine cables linking Nagasaki with Shanghai and Vladivostok as well as a cable between Nagasaki and Yokohama. In 1871 the Great Northern completed the first two cables, thus bringing Japan into the expanding worldwide telecommunications network.

From its humble start, Japan's domestic telegraph network expanded at an impressive pace in the early 1870s. The 1,340-kilometer Tokyo-Nagasaki trunk line was completed in 1873, followed two years later by the completion of the 825-kilometer Tokyo-Aomori trunk line. Telegraph construction also went on in Hokkaido and Kyushu simultaneously. It took only six years for the Meiji government to link all major Japanese cities, from Sapporo in the north to Kumamoto in the south, in a nationwide telegraphic communication network. In 1875 Japan's total telegraph lines had reached 1,760 li (6,912 kilometers or 4,294 miles), carrying 612,000 telegrams a year.[7]

The great promise of telegraphic communication was obvious to many Japanese, both inside and outside the government. Itô Hirobumi, then in charge of the Ministry of Engineering, which oversaw the telegraph construction, memorialized to the Meiji emperor in 1875 that "telegraph lines are most effective in facilitating public administration and promoting human intelligence. Now that landlines extend from Tokyo to both Nagasaki and Aomori, from Hakodate to Otaru in the north, even people in remote areas can benefit from our prosperity."[8] Itô's enthusiasm was shared by Fukuzawa Yukichi, the leading Japanese intellectual on modernization issues. "When we think about the function of the telegraph," he observed with apparent excitement at the opening of the Tokyo Central Telegraph Office, "we can either say that the distance of 1,700 li has been reduced to naught, or that the body of the Japanese has been extended to all locales. Since there are also telegraph lines in foreign countries, not only Japan but the entire world will be shrunk and made more manageable."[9]

Modern technologies such as telecommunications did not just flow naturally into Japan. Like nearly all other modern enterprises, early telegraphy developed in Japan by relying on foreign technical know-how for everything from construction to maintenance. For instance, British engineers were hired for the construction of the very first 32-kilometer line between Yokohama and Tokyo.[10] The dependency was even more pronounced in international communications. In fact, Japan's very access to the international telecommunications network came with considerable costs. The new Meiji government had neither the financial resources to invest in the costly submarine cables, nor the technological ability to lay them. This gave the Great Northern considerable bargaining power. Although the Danes attempted to secure a monopoly on Japan's communication with the Asian mainland, they failed, partly due to the strong opposition of the British Foreign Office. As a result, the treaty signed in 1870 stipulated that the Great Northern "will not make any complaint even if the Japanese government gives permission to others to begin the same service." However, the company would be entitled to the same benefit if Japan granted more benefit to companies from other countries.[11] The treaty allowed the Great Northern to land cables and to establish a telegraph station in Nagasaki, which was to remain in operation under the Danes until 1940.

Seeking autonomy and pursuing expansion have been intertwined in modern Japan from the very beginning. Nowhere was this clearer than in Japan's relations with Korea in the last decades of the nineteenth century. Japan's first attempt to exert influence in communication beyond its shores was with Korea. The 1876 Treaty of Kangwha, following the incident when a Japanese navy vessel surveying Korea's waters was fired upon, allowed Japan to open a post office in the southern Korean port city of Pusan. As Sino-Japanese rivalry in Korea became a major issue in Meiji foreign policy, the need for swift communication with the peninsula grew more urgent. The only channel of communication between Korea and Japan was by ship to either Shimonoseki in western Japan directly or to the Chinese port cities of Tianjin or Chefoo, where the message could be sent by telegraph to Nagasaki via Shanghai. Neither alternative was satisfactory. In 1882 political tension in Korea grew into crisis when anti-Japanese sentiment led to the expulsion of the Japanese minister and an attack on its legation. A direct link to Korea became an absolute necessity.

Lacking the necessary funds—due to the financial retrenchment known as the Matsukata deflation—and, perhaps more importantly, the technology, the Japanese government had no choice but to ask the Great Northern to lay the first submarine telegraph cable across the Korean Strait in December 1884. As a result of further negotiations with the Korean government, Japan was allowed to build a short landline from the Great Northern's landing site to the Japanese settlement in Pusan, where Japan would establish a telegraph office. For the first time, Japanese-language telegrams could be received and sent outside the Japanese islands, an important advantage to the small but growing Japanese presence in Korea. To protect its control over this vital communication link, the Japanese government secured a 25-year monopoly during which time Korea would not build competing telegraph lines itself or allow a foreign government or company to do the same.[12] In this way, Japan made innovative use of the "advanced West" in facilitating its own expansion in Asia.

All of this came at a price to Japan's own autonomy, however. In exchange for Great Northern's timely service, the Japanese government granted the Danish company a 20-year monopoly over Japan's international cable communications. In return, the Great Northern agreed to add new cables between Nagasaki and Shanghai and Nagasaki and Vladivostok. In addition, the company promised not to raise cable rates without the consent of the Japanese government and ensured that all submarine cables would remain in working order.[13] The arrangement meant foreign control of Japan's overseas communication for many more years, with serious economic and strategic consequences. Japan's attempt to establish an independent link with Asia would encounter persistent opposition from the Danish firm. The deal was much criticized by many Japanese ever after, who blamed those nearsighted bureaucrats for having placed Japan's international communication under the yoke of Great Northern monopoly for decades to come.

Japan had little choice, though. Apart from the technological challenge, the cable between Japan and Korea was estimated to cost ¥300,000, which equaled a quarter of all expenditures on telegraph construction in Japan during the 14-year period between 1869 and 1882. Moreover, the benefit seemed worth its price. Telegraphic traffic between Japan and Korea increased from a trickle of 3,800 telegrams in 1884 to some 100,000 annually a decade later. By 1905, about 430,000 telegrams, 98 percent of which were in Japanese, were transmitted through the Korean cable. Before Japan's final annexation of Korea in 1910, the Korea traffic occupied 60 percent of Japan's total international traffic. It was all the more remarkable that nearly all of it traveled on a completely foreign-owned cable until in 1891, when the Japanese government purchased the portion between Nagasaki and Tsushima. During the Sino-Japanese War, the Japanese government sought to buy the entire cable, but the Great Northern refused. It was not until after Japan annexed Korea that it finally was able to complete its control by purchasing the remainder of the cable.[14]

SUBMARINE CABLES AND JAPAN'S COLONIAL EMPIRE

Thanks to its victory over China in 1895, Japan was handsomely rewarded with concessions and indemnities. Japan acquired its first overseas colony, Taiwan, and with it an embryonic colonial empire. At the very inception of Japan's colonial empire, Japanese leaders recognized communications facilities as important infrastructure. No sooner had the two governments concluded the Shimonoseki Treaty than Japan's War Ministry set up a provisional agency for constructing telegraph facilities and navigation beacons in the new colony. Headed by General Baron Kodama Gentarô, the agency was charged with the establishment of a telegraphic link between the home islands and Japan's first overseas territory, lying some 2,000 miles across the ocean in the south.

This was not the first telegraph link to Taiwan. While Taiwan was still a remote province in the Chinese empire, it had become connected with mainland China by submarine cable in 1887, at the suggestion of a reform-minded Chinese official. The Taipei (Danshui)-Fuzhou (Chuanshishan) line was completed under the supervision of foreign engineers. The Chinese government was the main beneficiary of the telegraphic link to Taiwan, but it was also used by British and German merchants in the sugar and tea business, and even by some Chinese merchants from Shanghai.[15] As there were no direct telegraph links between Taiwan and Japan, the Japanese expedition forces sent to take over the island had to communicate with Tokyo by way of this link and via Chinese cities such as Fuzhou and Shanghai.

Its resounding military victory over China notwithstanding, Japan had just embarked upon its own industrialization and was yet to be initiated into the business of submarine telegraphy. It lacked the capability of either manufacturing or laying submarine cables. The large sum of indemnity received from China, however, made it possible to start several ambitious projects in telecommunications expansion, including the submarine cable link to Taiwan. Initially, opinions were divided within Japan on whether or not such an unprecedented project should be supervised by an experienced British engineer. Among those who objected were two engineers of the Ministry of Communications (MOC), who argued that the proposed British supervision would make Japan forever dependent on foreign skills in future submarine cable expansions. Their views won the support of General Kodama Gentarô and prevailed. As a result, Japan ordered its first cable laying ship, *Okinawa*, from a shipyard in Britain in addition to submarine telegraph cables. Laying of the cable, which began in July 1896 and was completed a year later, was carried out entirely by Japanese hands.[16] In his report to the war minister upon completion, General Kodama praised Japanese technicians for overcoming adversities ranging from rough ocean and terrain (landlines were built in hard-to-reach mountains for security reasons—to be out of sight of ships) as well as interruption by "local bandits" in Taiwan.[17]

The significance of the completion of the Taiwan-Japan submarine telegraph cable cannot be overemphasized. Technologically, the success of the first major modern telecommunications project outside Japan proper marked the country's ascendance as a world-class player in submarine telegraphic communications. The first cable ship acquired for this project was to play an active role during the next several decades; it gave Japan valuable experience in laying submarine cable in open seas. Warfare and colonial expansion benefited from the progress in telecommunications technologies; they also stimulated further technological development.[18]

Rapid means of communication between Japan and its first overseas colony seemed to favor adopting the French model of colonial assimilation. Hara Takashi, the future prime minister who served in the Taiwan Affairs

Bureau in 1895, listed relative geographical proximity and ethnic and cultural similarity among his reasons for integrating Taiwan into the homeland. He further noted that even the physical distance would be "shortened" with the completion of the submarine cables and the development of shipping routes.[19]

There were additional benefits as well. The exact length of submarine cables ordered from the British manufacturer was a military secret, but with remarkable foresight, General Kodama had placed an order for 1,648 nautical miles of submarine cables to cover 1,045 nautical mile route between Japan and Taiwan. As a result, the army was left with some 600 nautical miles of submarine cables as strategic reserves. Entrusted to the MOC for storage, this strategic cable reserve would prove to be a most critical asset less than a decade later. When tensions with Russia on the Korean Peninsula began to mount, this spare cable was put to use immediately for a submarine telegraph link between Korea and the Liaodong Peninsula.[20]

As soon as the Russo-Japanese War broke out in 1904, Japanese troops quickly seized control of Korea's telegraph network, despite Seoul's pledge of neutrality.[21] Japan's defeat of Russia in the war assured its status as the unchallenged power in postwar Korea and a major power in East Asia. Korea was made a Japanese protectorate, and Japan took control of Korea's telegraph, telephone, and mail systems. Ikeda Jûsaburô, an experienced MOC bureaucrat then in charge of the Tokyo Post and Telegraph Office, was dispatched with a team of Japanese technicians and officials to take over Korea's communications facilities. The Koreans put up last-ditch resistance to the Japanese attempt, but Ikeda succeeded after applying considerable coercion, by his own admission.[22] In the meantime Japan negotiated with the Great Northern Telegraph Company and acquired the remaining portion of the submarine cable in the Korean Strait. As Japan assumed complete control of the telecommunications system in Korea and made some improvements, it also brought about several reductions of telegram rates between Japan and Korea.[23]

Japan's annexation of Korea in 1910 allowed its strategists to cast their eyes from the peninsula to the continent. Ironically, the war with Russia exposed serious weakness in Japan's overseas communications. Not surprisingly, it was the military that raised the greatest alarm over Japan's overseas telegraphy connections. Some in the government felt that those wartime military lines should be withdrawn once the conflict was over, for their maintenance was costly and their value seemed negligible, not to mention diplomatic complications. This view, as the military saw it, amounted to ruining the "hundred-year blueprint" of the country:

> All countries of the world consider autonomy in communication a basic tenet in national defense. All aspire to possess their own communication systems. In the early years of Meiji, Japan gave in to the Great Northern Company that represented the Russian interests, and till today Japan has been unable to break away from its grips. At the present, with the newly established Anglo-Japanese Alliance, Japan is looked upon as a formidable Power. However, it could not even construct a submarine cable for the purpose of joint military operations. Isn't it a disgrace to the reputation of Japan? Leaving aside the issue of reputation, anyone who knows of the attitude of the Great Northern in the course of this war must realize the precarious position of Japan's overseas communication. Both land and underwater telegraph lines set up during the war gradually expanded with the progress of military operations, and the magnitude of its organization is truly spectacular in East Asia. However, this is nothing more than a temporary phenomenon based on belligerent rights. Whether or not they can be maintained in peacetime depends on the resolution of Great Northern's privileges.[24]

It was against such a background of geopolitical change and new cable development that for the first time the Japanese government began to formulate a comprehensive international telegraph policy consistent with Japan's new objectives in East Asia and its industrial strength.

Acutely aware of the presence of Western interests in neighboring China under various treaty protections, the Japanese government saw the monopolies of British and Danish cable companies as the main obstacle. The fact that Great Northern's ten-year monopoly was to expire at the end of 1912 provided a perfect opportunity for wresting control. In early 1912 the MOC set up a Committee for Investigating Overseas Telegraphy (Taigai Denshin Jimu Chôsakai). The Committee called for a general reduction of international cable rates, recommended setting up a new link with Russia, and proposed reopening the Taiwan-Fuzhou cable for international traffic. The centerpiece of the new policy, however, was an independent telegraphic route to China.[25] In fact, setting up Japan's own cable link to Shanghai, considered the international hub of news, finance, and commerce in the Far East, had been Japan's objective since the dawn of its international communications. Terashima Munenori, regarded by many as the "father of telecommunication" in Japan, was an early advocate of laying such a cable when he was still a hot-blooded young man in early Meiji.[26] Needless to say, Japan at that time possessed neither technological nor financial capacity to do so.

The overthrow of the Qing dynasty in 1911 in China and subsequent political instability added a new urgency for such a cable. The head of the

Japanese Chamber of Commerce in Shanghai petitioned the Japanese consul general the following year to install wireless sets in ships sailing between Japan and Shanghai. This was partly due to the political uncertainties in China, which required additional routes of communications with Japan. Moreover, installation of a wireless system would pressure the Great Northern to reconsider its policies. The company had held a monopoly over telegraphic communications between Japan and China and acted with arrogance, and as a consequence the Japanese residents had suffered for many years.[27] The military in Japan also demanded that a separate Japanese cable to China be given top priority in upcoming negotiations. Minister of Communication Gotô Shimpei, well known for his activities in the colonial administration in Taiwan and later in the railway management in Manchuria, agreed: "The trend among the great powers was the extension of telegraph network all over the world, and in particular the attempts to open telegraphic links with the great market of China."[28]

As the Japanese officials were fully aware, the greatest obstacle to Japan's aspiration was the Great Northern, which would be threatened by a potential loss of its monopoly. Captain J. J. Bahnson, the company's general manager for the Far East, known among the Japanese for his arrogance, proved to be a tough negotiator, steadfastly refusing to relinquish the company's permanent monopoly over Japan's overseas telegraphic communication. He maintained that existing cables were sufficient and even offered to handle Japanese-language telegrams. He eventually, however, gave in to Japan's demand to lay a new cable between Shanghai and Nagasaki, after Japan promised the new cable would not cause financial damage to the company. The new Shanghai cable would handle Japanese-language telegrams between Japan and Shanghai as well as European-language government telegrams between Japan and China. In return, Japan agreed to set up a joint purse and share with the Great Northern the revenue of all telegraphic traffic between Japan and China until 1930, turning over as much as 64.5 percent of the revenue from the new Shanghai cable. During the 16-year period, this payment alone would amount to over ¥25 million.[29] Once again, financial sacrifice was the price for strategic advantage.

Once the Great Northern gave in, negotiations with Chinese representatives in Tokyo proved much easier. In exchange for China's consent to the new Japanese cable, Japan granted all Chinese government telegrams a 50 percent discount and agreed to pay China a five-cent-per-word terminal rate for all telegrams sent over the cable. In a secret agreement, both governments pledged that they would not extend privileges in telecommunications between the two countries to a third party without first consulting each other. In particular, the Japanese insisted that China not extend landing rights to the Great Northern after their expiration in 1930. Interestingly, the definition of "electric communications" caused considerable disagreement between the Japanese and Chinese negotiators. The Chinese preferred to specify "submarine telegraphy" so as to exclude wireless and telephone from the agreement. Since they had lost much of their submarine cables' sovereignty to foreign interests, the Chinese were determined not to repeat the same mistakes with the wireless facilities. The Japanese negotiator insisted that Japan had no such intentions, but refused to put it in writing. In the end, the Japanese prevailed.[30]

The first completely Japan-owned Nagasaki-Shanghai cable opened to public service on the first day of 1915, just as World War I began to engulf the entire continent of Europe. The inauguration of Japanese-language telegraphic communication between Japan and China was just in time for the rapid expansion of Japan's presence on the continent during the European War. By the early 1930s, more than half of the telegrams exchanged between Japan and China proper went through this Shanghai cable, averaging over a thousand telegrams daily.[31]

In a report submitted to the prime minister on Japan's international telegraphic communications shortly after the agreement with China, Minister of Communications Motoda Hajime proudly enumerated major rights and concessions (riken) that Japan had recently acquired. The new cable to Shanghai, despite its cost totaling ¥1.62 million, "had grave implications in terms of military affairs, commerce, and diplomacy. It had been an age-old objective for Japan." As Minister Motoda put it, the agreement with China would serve as "a major guarantee for Japan and a check on China's telegraphs" (tai-Shi denshin no seichû).[32] Given Japan's own objective in augmenting its influence over China's communications, such a clause in effect paved the way for Japan's ascendance in that field. In addition, Japan acquired a telegraphic link with Russia via Sakhalin and brought about sharp reductions in worldwide telegram rates on an average of 26 percent. Taken together, Motoda emphasized, these new gains would become the foundation of Japan's future international telecommunications policy.[33]

In addition, following Japan's attack on the German stronghold of Qingdao in the Shandong Peninsula, Japan's cable-laying capabilities enabled it to sever Germany's Shanghai-Qingdao cable and in May 1915 connect it between Qingdao and its own naval base in Sasebo.[34] By the end of World War I, Japan finally managed to succeed in the West's own game of informal imperialism and gained a significant presence in the telecommunication field in China.

Figure 11.1
Japanese submarine cables in East Asia (1915).

A Cable Network for the New Imperium

Just as Japan's efforts to expand its submarine cable routes began to bear fruit, new technological developments appeared to be altering the status quo. The government came to view the advent of wireless communications as a heaven-sent blessing to its effort to end foreign dominance of its international cable routes. Traffic over wireless facilities operated by Japan would reduce payments to the foreign cable company. Compared with submarine cables, wireless communication required less setup cost and would encounter less difficulty than securing landing rights in foreign territories and was thus deemed most desirable in establishing direct communications with countries that were of great importance to Japan.[35]

Given these new promises, the government began studying measures to boost Japan's communications capability through wireless. Army, navy, and communications ministries conducted joint investigations in 1924 on high-power wireless for overseas communications and reaffirmed the urgent need to "quickly establish Japan's superior position in international communications in East Asia." The committee proposed that wireless telegraph be given precedence over cable in the future expansion of Japan's international telecommunications facilities and called for direct wireless communications with foreign countries without relay in third countries.[36] By the decade's end, Japan had secured a place in international wireless communication through institutional innovations such as the Japan Wireless Telegraph Company and later the International Telephone Company.

Even with the new technology, however, the goal of establishing a dominant position on the continent remained elusive. Japan's wireless venture in China failed miserably, due to intensified international wireless rivalry as well as rising Chinese nationalism. China, moreover, began to harness the new technology to enhance its own communication autonomy. It was after the Manchurian Incident of 1931 that technological advances and the new imperial agenda combined to bring about another major reformulation in Japan's communications and information policy. In the mid-1930s, Japan embarked on an ambitious project of establishing an expanded telecommunications network for the portions of East Asia that had come under its influence.

Full-scale conflict broke out between Japan and China in the summer of 1937. By late 1938 Japan had occupied most of China's coastal areas and established several regional telecommunications companies in those areas. Japan now considered building strong communication ties between Japan, Manchukuo, and China a top priority. MOC officials and engineers were busy working on details of the blueprint. In September 1938 the MOC set up the Research Group on East Asian Long Distance Telecommunications Network. Led largely by section chiefs, the group mobilized nearly all of its manpower to conduct a wide range of studies, including detailed estimate of cost, demand, and revenue.[37] After deliberation and consultation, the telecommunications committee issued its first policy recommendation in January of 1939. In its recommendations, a newly established interministry Telecommunication Committee called for immediate planning of a secure and strong communications cable that would link political, military, economic, and cultural metropolitan centers in Japan, Manchukuo, and China that were most closely tied to the realization of Japan's continental policy.

The first stage of the grand overseas telecommunications expansion was the construction of a cable network that would function as "the great artery of Japan-Manchukuo-China communication" (see figure 11.2). The network would consist of two major cable routes with four branches. They would extend first on land from Tokyo to Fukuoka and Nagasaki via Nagoya, Osaka, and Shikoku. From Fukuoka, it would cross the Korean Strait and eventually reach Tianjin and Beijing via Pusan, Keijo, Andong, Mukden, Shanhaiguan (in north China), while one branch would extend from Keijo to Nanyo via Wonson and Jozu (in northeast Korea). From Nagasaki, the second cable would extend by way of Cheju Island to Shanghai, and from there one branch would farther extend on land to Nanjing, with another to Takao (present-day Kaoshiung) in southern Taiwan via Taihoku (present-day Taipei). In the future, this East Asian telecommunications network would extend farther—from northern Korea to the Soviet boarder in the north and from Taiwan to south China and Southeast Asia in the south.[38] In between it would cover all important areas in central and north China. The construction of this network thus fully embodied the visions of the inner core area of what eventually became the Greater East Asia Co-prosperity Sphere.

The plans for a cable-based imperial telecommunications network in East Asia reflected a new emphasis in Japan's communication strategy. Ever since the Meiji era, autonomy in international communication had been one of the most basic tenets of Japan's foreign telecommunications policy.[39] Bound by treaties that had granted the lion's share of international telecommunication traffic to the Great Northern, the joint army-navy-MOC investigation of 1924 had recommended wireless telegraph over cable in the future expansion of Japan's international telecommunications.[40] Japan had subsequently hastened the pace of opening direct wireless communications with many countries in Asia, Europe, and the Americas. By the early 1930s wireless facilities were carrying about half of Japan's international telecommunications traffic.[41] In his radio speech at the beginning of 1936, a leading MOC official still referred to a coming age of wireless.

While the new priority given to cable technology did not mean abandoning wireless, the shift in budgetary emphasis raised a number of questions from skeptics. During the Imperial Diet deliberation in early 1939, for instance, a member of the House of Peers asked whether, given the continuing improvement in wireless and other new technological inventions, such as television, such an ambitiously huge project based on costly cables might soon lose its value.[42] Indeed, after tirelessly championing the wireless for over a decade, why did the MOC suddenly have a change of heart? Advocates for a cable network had to present a convincing case, however. The answer lies in the fact that cable was better suited to Japan's imperial network.

Figure 11.2
A blueprint of the East Asian cable communications network.

First, communication capacity was given much consideration. As Japan continued to expand and consolidate its political, military, and economic presence on the Asian continent, a sharp increase in telecommunication traffic was anticipated. Such traffic would increase with new industrialization plans and further Japanese migrations. Telecommunication Bureau Director Tamura Kenjirô pointed out that, owing to the limitation of suitable radio frequencies, wireless communication could no longer meet the requirements of the Greater Communications Network in East Asia. As of June 1932, of the total 1,878 shortwave frequencies used worldwide, Japan had 212, divided between the MOC (124) and the military (88). Writing in a trade magazine, Tamura predicted that 47 telegraph circuits and 28 telephone circuits would be required by 1944 to link Japan, Manchukuo, and north China.[43] In view of its great importance, the Japan-Manchukuo route alone, he told a committee of the Imperial Diet, could conceivably require hundreds of channels compared to the few that then existed. The new NLC multichannel network promised greater capacity for the anticipated voluminous traffic between Japan and its adjacent imperial sphere of influence. Given the limits of wireless frequencies available, cable appeared the clear winner. Moreover, the limited frequencies available to Japan had to be reserved mainly for communication with more distant foreign countries in Europe or the Americas.

In the meantime, the government placed greater emphasis on communication secrecy. While telegraph could rely on encryption codes, as were widely used at the time, telephony posed a greater technical challenge to security.[44] As most long-distance telephone links were by wireless, Japan had already applied early scrambler technology to its most sensitive circuits between Japan and Manchukuo. Tamura Kenjirô noted that since Japan, Manchukuo, and China were bound by a special relationship, military, political, and diplomatic communication among them required special measures of protection against interception. Given the advances in code-breaking techniques, Tamura told the committee, wireless communication ran the risk of being easily intercepted and deciphered by the enemy. In contrast, communications by cables would have a decided security advantage over wireless.[45]

Finally, communications stability and safety favored the cable solution. Although improvements in wireless technology continued to be made, the perennial problem of atmospheric interference with wireless communications still remained. The problem of static was severe enough to interrupt wireless communication at times. By contrast, a network of underground or underwater metal-shielded cables would be entirely stable and safe. MOC officials were aware of the fact that much of the land-based network would be in territories outside Japan's formal empire, but sabotage could be eliminated with the underground cable. Even with broadcasting, which was typically associated with wireless, cable technology had its advantages: the greater capacity of imperial trunk cables could be used to facilitate program exchanges among Japan and other parts of the empire regularly. Local wired networks would provide advantage over wireless for the sake of protection against enemy air raid.[46] Thus cable lines would provide the greater capacity, security, and stability required of Japan's imperial communications network.

Of all the routes proposed by the Telecommunications Committee, two were entirely new compared to the previous plans. The second Japan-Manchukuo cable, which extended from Keijo to Mudanjiang via the east coast of Korea, was for the purpose of strengthening national defense along the Soviet border as well as exploiting the natural resources in northeast Manchuria and northern Korea. Given the perceived importance of Japan-Manchukuo ties, this cable would serve as a backup route to the first one. The proposed telephone cable between Nagasaki and Shanghai was considered necessary because Shanghai was the base of Japan's activities in central China in war and afterward. Yet there was no secure telephone connection apart from a wireless route. Moreover, the Shanghai-Nagasaki telephone cable would be further extended to Taiwan and thus was the first stage of southward expansion of the entire network. The combined cost of these two cables would reach ¥45 million, over one-third of the total estimate of the entire East Asian cable network.[47]

The choice of the more expensive cable for the imperial network therefore was not accidental but can be said to reflect an important aspect of Japan's empire-building project of the 1930s as a whole. Although much higher in construction and maintenance costs, the long-distance cable network for East Asia would be more durable and therefore more appropriate for Japan's lasting imperial enterprise.

Imperial Ambitions and Realities

The outbreak of the war in Europe in 1939 opened new opportunities for Japan in Southeast Asia. In mid-July 1940 the MOC prepared several confidential plans on the feasibility of setting up a regional communications network in Southeast Asia that would be later be incorporated into the Japan-Manchukuo-China network. Despite the important location of the region, one report pointed out, telecommunications facilities in the area were underdeveloped compared to areas under Japan's occupation, such as north China or Inner Mongolia.

As a solution to such problems, Japan was to construct a new submarine cable from Japan and China and connect it with the internal network that already existed in Southeast Asia. In support of this proposition, the MOC report specifically singled out the "sturdy and highly efficient Japanese cable technology as used in Japan-Manchukuo-China lines" as a key factor in expanding telephone, facsimile, and broadcast relay capabilities in Southeast Asia. There was apparent confidence that Japan's technology was to prevail over telecommunications facilities in the Dutch East Indies, the most advanced in the region.[48] Given the political conditions in French Indochina and the Dutch East Indies, the MOC outline suggested measures that would induce their respective governments to privatize both domestic and international telecommunications to make it easier for them to "become dependent on Japan's communications influences." Specifically, Japan should strive to possess "strong influence" over technical and design matters that were the basis of telecommunication facilities.[49]

By late July of 1940, MOC officials had drafted perhaps the most ambitious and comprehensive plan ever made for telecommunications expansion in Asia. Over the next five years, the plan called for the completion of ten major trunk cables in addition to a large number of wireless facilities in order to form a so-called East Asian Stability Sphere Telecommunications Network. The geographical scope would encompass Northeast Asia, Southeast Asia, and South Pacific Mandate Islands, reaching as far as Australia. The total cost, which also covered the construction of three new cable ships, would be a staggering ¥600 million—an almost fivefold increase from the estimated cost of the East Asian Trunk Cable Network drawn up a year and a half before. In terms of technology, nonloaded cables would be used on land in Manchuria, China proper, and Indochina, but two underwater telephone cables and three underwater telegraph cables (all unloaded) would be laid in Southeast Asia. In addition, all major cities in Southeast Asia and Australia would be connected by a powerful wireless system.[50]

With an eye to extending future operations in Southeast Asia, the navy also called for building new submarine cables linking Japan and the southern region. Of all the submarine cables proposed, the navy considered two to be of greatest urgency: one cable would extend from Nagasaki to Shanghai, Taipei, Takao, Haikou, Hanoi, Bangkok, and Batavia; the other would connect Tokyo to Saipan, Yap, Palau, Menadao, Talakan, and Balikbanban. With these two new cables, the navy predicted, Japan would have secure communication routes to the region independent of Western cables or the often unpredictable wireless.[51] Relatively new in mapping communications expansion, however, the navy proposal looked modest in comparison to that of the MOC.

At the end of August, the Cabinet Planning Board discussed the MOC proposal. It came to the conclusion that of the proposed trunk cables, the South China Sea Western Circular Cable, which would run from Takao in southern Taiwan to Batavia in the Dutch East Indies via Canton, Hainan Island, and Saigon, was to be given the highest priority. A cable landing in Singapore, as suggested in the original proposal, was dropped to avoid complications with the British. Compared with other routes, the MOC noted,

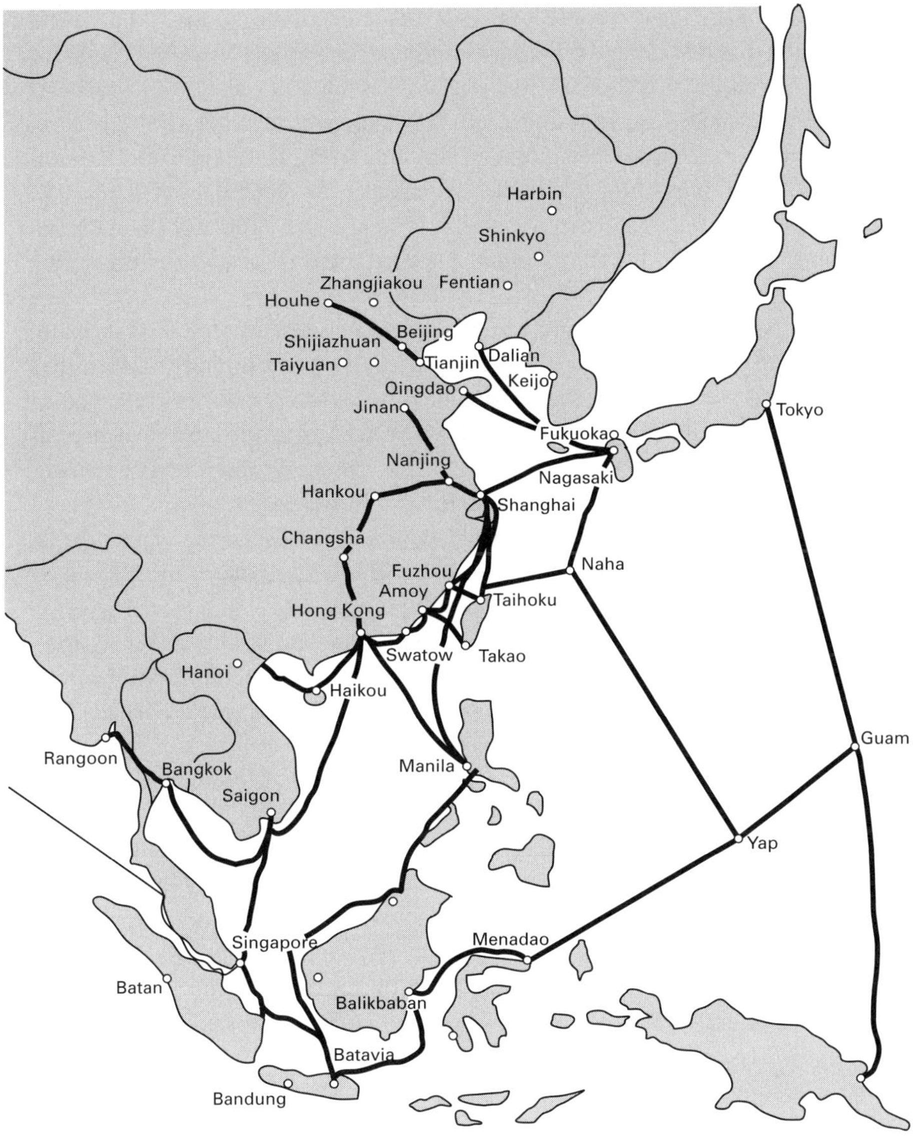

Figure11.3
Cable routes in the "East Asian Stability Sphere" (1940).

this cable was "most feasible based on the current situations." The MOC predicted that this cable would draw traffic not only from Japan but also from China, given the presence of large overseas Chinese population in Southeast Asia. By its calculation, in the first year this cable would generate ¥12 million for Japan from telegram, telephone, telephotography, as well as from leases on telephone circuits. The MOC based its calculation on actual traffic in 1939 as well as the annual increase rate on the Japan-Manchukuo and Japan-China circuits.[52] In November the Cabinet approved the final plan with only minor changes.

In September 1940 Japan signed the Tripartite Pact with Nazi Germany and Italy, a treaty stipulating that the signatories would support one another militarily if any one of them was attacked by a country not yet involved in the war. Soon afterward, Japanese troops moved into the northern part of French Indochina, partly in an effort to cut off supply lines to the Chiang Kai-shek government in China. There was, however, no reason to expect that other European colonial powers would yield to Japan's demands. In fact, Japan's action hardened the response of the United States and Britain to curtail Japan's ambition in Southeast Asia.

Even before the war broke out in the Pacific and while the MOC continued to produce ambitious expansion plans, Japan's telecommunications in East Asia were already beginning to show signs of serious strain. This was best illustrated by the failure of the Nagasaki-Shanghai underwater telephone cable, one of the key components of the East Asian telecommunications network.

Shortly after the outbreak of the war in China and while the Japan-Manchukuo long-distance cable was still in progress, the MOC proposed a 1,102-kilometer submarine telephone cable linking Japan and Shanghai via Cheju Island. The navy was particularly interested, as it promised to protect conversations from being intercepted by the enemy. The initial design, unveiled in early 1938, featured eight voice circuits between Nagasaki and Shanghai. In addition, there would be one broadcast relay channel, allowing cable transmission of radio programs. In October 1938 the government decided that the planned cable would be extended from Shanghai to Taiwan, as the first step to further expansion to Southeast Asia. Together with its broadcasting channel, total voice channels would be increased from eight to ten. Preliminary exploration began in July 1938. Together with the Japan-Manchukuo long-distance cable, which was being extended southward to Beijing and Tianjin, the planned Nagasaki-Shanghai cable would complete the core of the telecommunications cable network in Northeast Asia as well as the first stage of the network expansion in Southeast Asia. By then, Japan had already built a 3,760-ton cableship, *Tôyô* (Eastern Seas), to replace the aging *Okinawa*. The age of submarine cable expansion seemed ready to start.[53]

The Nagasaki-Shanghai telephone cable was to be one of the most ambitious submarine cable projects ever launched by the MOC. From the beginning, this was destined to be a most costly project. The budget for investigation and construction of one relay station alone was set at ¥1 million. The project also posed an unprecedented technological challenge in submarine telephony, especially given the 454-kilometer stretch of ocean between Cheju Island and Shanghai. The idea of floating relay stations was considered, but in the end it was deemed better to construct a relay station on a no-man's island halfway between that would also serve as a beacon in the busy shipping lane between Japan and central China. Because coaxial cable was too new a technology, the MOC decided to use lead-shielded submarine cables, which had normal depth limit of 300 meters. All three major cable manufacturers in Japan devoted considerable effort to building underwater telephone cables that could sustain high pressure and support quality voice transmission. To make amplification possible in such a long submarine cable, Japanese engineers proposed inserting amplifiers inside the cable, using small but extremely stable vacuum tubes with electricity in direct current provided from both ends of the cable.

Actual construction proved to be more difficult than expected, while cables produced by the three leading cable manufacturers in Japan failed to meet the requirement. By early 1940, the MOC had once again to revise the early design for the cable. As Japan moved to a war economy under national mobilization plans, securing required material—1,200 tons of copper and 5,000 tons of lead for the Shanghai-Nagasaki cable alone—and continuing with costly research became unrealistic goals. The project was finally abandoned after the outbreak of the Pacific War.[54]

The failure of the project had important consequences for Japan's plans to expand the telecommunications network southward. When the cabinet approved the plan to first build a submarine telephone cable from Taiwan to Batavia (South China Sea Western Circular Cable) in late 1940, the Nagasaki-Shanghai project was already in serious trouble. It was highly doubtful—even if the Dutch authorities had agreed to Japan's demand to land a new cable to the Dutch East Indies—that Japan would have been able to produce enough new cables for the project, at least in the short run.

Japan's telecommunications network did enjoy a brief boost after the outbreak of war in the Pacific. At its height, Japan's submarine cable network extended to 15,000 nautical miles. Even this, however, turned out to be a hollow victory. Most of the facilities that Japan acquired in Southeast Asia, as it turned out, were out of operation or required enormous resources for maintenance. All Japan was able to produce in the region was the external shielding on 300 nautical miles of cable core already manufactured.

Japan did increase its own production of gutta-percha submarine cables at home, thanks to the gutta-percha from occupied lands. The ebb and flow of Japan's capability in cable manufacturing is clearly reflected in the production of gutta-percha submarine cables at home. Gutta-percha cable produced for the MOC, to be used for its planned expansion, reached a peak in 1940, after which cable for the military—the navy for the most part—increased sharply and took the largest share of production. After 1943, however, total production began to drop as Japan's economy suffered increasingly from shortage of material. This is all the more ironic in that its occupation of Southeast Asia for the first time gave Japan direct access to gutta-percha—the raw material indispensable to making submarine cables. Japan's own effort at refining gutta-percha failed in Singapore. Despite requests from the military, even attempts to produce gum-coated wires locally did not succeed.[55] As Japanese discovered, although there was an overabundance of such material as gutta-percha, aluminum, and tin in Southeast Asia, Japan was short of other materials, such as copper and nickel, which were indispensable in producing telecommunications equipment.[56]

Owing to the inadequacy of submarine cables that remained in operation, Japan had to rely primarily on wireless communication with Southeast Asia. Reliance on the wireless created its own problems, however. Communications security became of increasing concern to the military, given the unprotected nature of wireless. To prevent security breaks, the Tokyo International Communications Promotion Society (Tokyo kokusai tsûshin shinkôkai) hosted a meeting in mid-1943 specifically dealing with security matters, bringing together business executives in the field of finance, general trade, shipping, and newspapers.[57] Nevertheless, Japan's increasing reliance on the wireless within its wartime empire gave the Allies a decisive advantage.

Conclusion

Japan's initial encounter with and later return to the submarine cables fit a general pattern of modern imperialism. As this paper has demonstrated, the development of submarine telegraphy in modern Japan was largely coterminous with the country's emergence as a colonial empire. Therefore, while submarine telegraphic cables were a technology of consolidation for the British Empire, as suggested by Daniel Headrick, they functioned largely as a

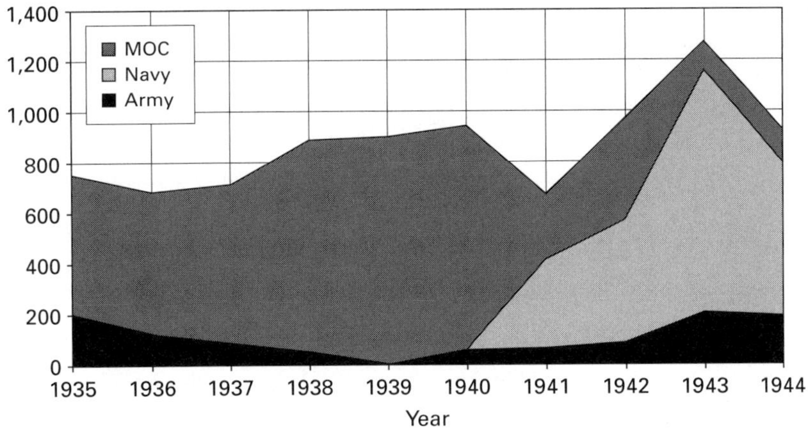

Figure 11.4
Length of gutta-percha submarine cable produced for the Japanese army, navy, and Ministry of Communications during the war years (in kilometers).

technology of expansion for Meiji Japan.[58] By the late 1930s, when the new East Asian long-distance cable network came to be seen as the artery of the new imperium, it also became a technology of consolidation. However, as historian Peter Duus has pointed out in his essay on Japan's wartime empire, "Instead of pursuing a manageable and minimal expansionist agenda as the Meiji oligarchs had, the leaders in the 1930s and 1940s pursued a maximal and unmanageable one." Their confidence, I argue, derived in large part from the earlier success in submarine and land cable networks in Northeast Asia.[59] In this sense, the technology of telecommunications enabled Japan to imagine its wartime empire that would eventually encompass a space nine times the size of its formal empire in 1930.

Japan's initial choice of cable over wireless for this new imperial network is also interesting. In his classic work on communication and empire, Harold A. Innis pointed out the difference between the technology that favors time over space and the technology that favors space over time. While Innis confined his analysis to communication before the age of the telegraph, it seems that submarine cables belong to the former category, while wireless belongs to the latter. As the Japanese government sought to justify the choice of cable over wireless in its ambitious East Asian network, the durability and capacity of the cable were emphasized as appropriate for Japan's long-term imperial project. Ironically, Japan's spatial expansion during the war exceeded its own capability to restore and expand the cable network, so that in the end it had to resort to wireless. In this sense, the choice of communication technology mirrored the fate of the wartime empire.

Notes

1. Preface to Nippon Denshin Denwa Kôsha Kaiteisen Shisetsu Jimusho, *Kaiteisen hyakunen no ayumi* (One hundred years of the submarine cables) (Tokyo: Denki tsûshin kyôkai, 1971).

2. James Carey, *Communication as Culture* (Boston: Unwin Hyman, 1989), 212.

3. Daniel Headrick, *The Tools of Empire: Technology and European Imperialism in the Nineteenth Century* (New York: Oxford University Press, 1981), 160–164.

4. Ibid., 11–12.

5. Daniel R. Headrick, *The Tentacles of Progress: Technology Transfer in the Age of Imperialism, 1850–1940* (New York: Oxford University Press, 1988), 98. See also his *The Invisible Weapon: Telecommunication and International Politics, 1851–1945* (New York: Oxford University Press, 1991).

6. On the activities of these two companies in East Asia prior to World War I, see Jorma Ahvenainen, *The Far Eastern Telegraphs*, Annales Academiae Scientarum Fennicae, ser. B, 216 (Helsinki: Suomalainen Tiedeakatemia, 1981). See also Kurt Jacobsen's chapter in this book.

7. Wakai Noboru and Takahashi Yuzo, *Terekomu no yôake* (The dawn of telecommunications) (Tokyo: Denki Tsûshin Shinkôkai, 1994), 76–78.

8. Quoted in Watanabe Masami, *Nihon denshin denwa sôgyô shi* (A history of the founding of the telegraph and the telephone) (Tokyo: Hifumi Shobô, 1958), 71.

9. Fukuzawa Yukichi, "Denshin kaigyô no shukuji," reprinted in Wakai and Takahashi, *Terekomu no yôake*, 97.

10. Nihon Kagakushi Gakkai, comp., *Nippon kagaku gijutsushi taikei* (An outline of the history of science and technology in Japan) (Tokyo: Daiichi hôki, 1969), 53–55; Takahashi Zenshichi, *Oyatoi Gaikokujin: Tsushin* (The hired foreigners: Communications) (Tokyo: Kondo shuppan, 1969).

11. Nagashima Yôichi, "Taihoku Denshin Kaisha no Nihon shinshutsu to sono haikei," *Nihon Rekishi* 367 (1995): 77–92. The text of the treaty is reprinted in *Nippon kagaku gijutsushi taikei*, 48–50.

12. For a complete text, see Ishikawa Junkichi, comp., *Kokka sôdoin shi* (A history of national mobilization; hereafter *KSS*), vol. 3 (Fujisawa: Kankôkai, 1984), 717–720.

13. For a complete text of the agreement, see ibid., 3:696–701.

14. *Chôsen teishin jigyô* (Communication in Korea) (Keijo: Chôsen sotofuku teishinkyoku, 1938), 107–109, 133; *Kaiteisen hyakunen no ayumi*, 171; Ahvenainen, *The Far Eastern Telegraphs*, 186.

15. *Taiwan nenkan* (Taiwan Yearbook) (Taihoku, 1925), 184; Taiwan yinhang jinji yanjiushi, *Taiwan jiaotong shi* (A history of communications in Taiwan) (Taipei, 1955), 11–12.

16. Funatsu Shigetarô, "Okinawa-maru to taiyô kaiteisen," in Teishin Gaishi Kankôkai, *Teishin shiwa* (Historical tales of communications) (Tokyo: Denki tsûshin kyôkai, 1962), 1:277–283.

17. Kodama Gentarô, "Denshin tôhyô kôji konkai" (September 30, 1897), reprinted in *Kaiteisen hyakunen no ayumi*, 144–145.

18. For a stimulating essay on the general theme, see Kozo Yamamura, "Success Ill-gotten? The Role of Meiji Militarism in Japan's Technological Progress," *Journal of Economic History* 37, no. 1 (March 1977): 113–135.

19. Edward I-te Chen, "The Attempt to Integrate the Empire: Legal Perspectives," in *The Japanese Colonial Empire, 1895–1945*, ed. Ramon H. Myers and Mark R. Peattie (Princeton: Princeton University Press, 1984), 250–251.

20. In *Kaiteisen hyakunen no ayumi*, 156.

21. Peter Duus, *The Abacus and the Sword: The Japanese Penetration of Korea, 1895–1910* (Berkeley: University of California Press, 1995), 184–186.

22. See the recollection by Okamoto Keijirô at a roundtable discussion in 1935. Okamoto was one of the seven Japanese officials sent to Korea in May 1905 to take over its communications administration. Chôsen sôtofuku, *Teishin shui* (Anecdotes of communications) (Keijo: Chôsen teishin kyôkai, 1936), 10–16.

23. *Chôsen teishin jigyô*, 107–109, 133; *Kaiteisen hyakunen no ayumi*, 171.

24. "Sengo ni okeru teikoku denshin seisaku" (n.d.), 1.7.4.36, Japanese Ministry of Foreign Affairs Archives (hereafter JMFA Archives).

25. Hanaoka Kaoru, *Kaitei densen to Taiheiyô no hyakunen* (One hundred years of submarine cables and the Pacific Ocean) (Tokyo: Nittô Shuppansha, 1968), 174–175.

26. Ahvenainen, *The Far Eastern Telegraphs*, 187; Machida, "Taigai denshin ryakushi," *Denki tsûshin gyômu kenkyû* 19 (July 1951): 60.

27. "Nihon Shanhai kan kisen ni musen denshin sochi ni kansuru ken," Japanese Ministry of Communications Papers, First Series (hereafter MOC Papers I).

28. Memo by Gotô Shimpei, dated February 6, 1912, *KSS* 3:734–738; "Daihoku Denshin Kaisha ni ataetaru tokkyo kigen manryôgo ni okeru teikoku gaishin seisaku ni taisuru gunjijô no yôkyû," vice minister of communications to vice minister of foreign affairs, September 18, 1913, 1.7.4.36-1, vol. 1, JMFA Archives.

29. Minutes of these negotiations can be found in 1.7.4.36-3, vol. 1, JMFA. See also Jiaotongbu jiaotongshi bianchuan weiyuanhui, comp., *Jiaotong shi: Dianxin bian* (A history of communications: Telecommunications) (Nanjing, 1937) 1:173–177; *Kaiteisen hyakunen no ayumi*, 180–181. Matsunaga Tadao, "Shin'yu mudentai no fukkô to Nisshi kan denki tsûshin no konseki," *Teishin kyôkai zasshi* 370 (June 1939): 38 (hereafter *TKZ*).

30. Chinese Ministry of Foreign Affairs to Japanese Minister to China, January 13, 1912, *Jiaotong shi: Dianxin bian*, 1:177–178. See also memorandum of a meeting between Chinese

Foreign Minister (Lu Zhengxiang) and Japanese Minister (Ijuin), January 14, 1913, in Zhongyang yanjiuyuan jingdaishi yanjiusuo comp., *Zhong-Ri guanxi shiliao: Yudian hang yu yan lin jiaoshe, 1912–1916* (Documents on Sino-Japanese relations: Negotiations over post and telecommunications, navigation, fishery, and forestry) (Taipei: Zhongyang yanjiuyuan jingdaishi yanjiusuo, 1975) 2:52–53. For minutes of the negotiations in Tokyo, see 1.7.4.36-4, JMFA Archives.

31. Iino Takeo, "Tai-Ka denshin mondai no keii ni tsuite," *TKZ* 281 (January 1932): 45. The author, a Japanese Ministry of Communications official, reiterated that given Shanghai's role in influencing world opinion about the Far East, the absence of its own telegraph link to the city was a grave disadvantage for Japan's information policy.

32. "Taigai denshin mondai ni kansuru sho kyôyaku ketsuryô ni tsuki hôkoku," October 7, 1913, reproduced in *KSS* 3:740–744.

33. Ibid., 741–744; Hanaoka, *Kaitei densen to Taiheiyô no hyakunen*, 177.

34. *Kaiteisen hyakunen no ayumi*, 184–205.

35. Nomura Yoshio, "Taigai musen denshin no kakuchô ni tsuite," *TKZ* 332 (April 1936): 145–149; Kobayashi Takeji, "Nihon musen denshin kabushiki kaisha-ho no kaisei ni tsuite," *TKZ* 345 (May 1937): 44–49.

36. "Rikugun daijin, kaigun daijin, teishin daijin ni teishitsu taigai dai musen keikaku ni kansuru sanshô kankei kanri no chôsa hôkokusho" (received on August 12, 1924), MT 3.6.11.23, JMFA Archives.

37. "Tôa chôkyori denki tsûshinmô chôsakai no ken," (December 9, 1938), Japanese Ministry of Communications Papers, Second Series (hereafter MOC Papers II), 362.

38. Ibid.

39. Nomura Yoshio, "Taigai musen denshin no kakuchô ni tsuite," *TKZ* 332 (April 1936): 145–149; Kobayashi Takeji, "Nihon musen denshin kabushiki kaisha-hô no kaisei ni tsuite," *TKZ* 345 (May 1937): 44–49.

40. "Rikugun daijin, kaigun daijin, teishin daijin ni teishitsu taigai dai musen keikaku ni kansuru sanshô kankei kanri no chôsa hôkokusho," received on August 12, 1924, Microfilm MT 3.6.11.23, JMFA Archives.

41. Tamura Kenjirô, "Kokusai Denwa Kabushiki Kaisha no sôritsu kettei made," *TKZ* 292 (December 1932): 82–96. By one rough calculation, some 177 remained unused worldwide.

42. *Teikoku gikai kizokuin gijiroku* (Minutes of the Imperial Diet House of Peers) (March 20, 1939), 5–6.

43. Tamura Kenjirô, "Tôa denki tsûshin seisaku to Kokusai Denki Tsûshin Kabushiki Kaisha no kakuju ni tsuite," *TKZ* 369 (May 1939): 37.

44. One of the young engineers at the Central Laboratory of the giant South Manchurian Railway Company claimed to have invented a device for secret telephone conversations in 1936. "Himitsu denwa no hatsumei," *Manshû Nichinichi*, June 3, 1936.

45. *Teikoku gikai kizokuin gijiroku*, 383. Much has been written about the utter failure of Japan's communication security, but the bulk of communication traffic intercepted by the Allies was diplomatic and military correspondence with embassies located outside this network. Japanese engineers did begin developing scramblers used on wireless telephone to ensure its secrecy, but in general they were not satisfactory.

46. On wire-based broadcast, see Shinohara Noboru, "Daitôa sensô to denki tsûshin gijutsu," *TKZ* 402 (February 1942): 5–6.

47. Denmukyoku keikakuka, "Daini Nichi-Man denwa keburu oyobi Nagasaki-Shanhai kaitei denwa keburu shisetsu gaiyô" (n.d.), MOC Papers II-693.

48. "Ran'in o chûshin tosuru Tônan Ajia chihô no tsûshinmô seibi ni kansuru ken," July 19, 1940, MOC Papers II-244.

49. "Tai Nanpô tsûshin seisaku yôkô," July 30, 1940, MOC Papers II-244.

50. "Tôa anteiken tsûshinmô keikakusho," July 20, 1940, MOC Papers II-244.

51. Vice Navy Minister, "Tsûshin shisetsu no seibi kakujû ni kansuru kyôgi," September 4, 1940, MOC Papers II-244.

52. See appendix to "Tai-Nanpô tsûshin seisaku yôkô," July 30, 1940, MOC Papers II-244.

53. *Kaitensen hyakunen no ayumi*, 284–304. When the cable ship was designed in 1936, some had proposed to build a larger one at 6,000–7,000 tons in anticipation of the future expansion of Japan's submarine cables network. The tonnage was eventually reduced so that the ship could also operate in shallow waters around Japan. However, the fuel capacity (burning coal instead of oil in anticipation of the latter's shortage in time of war) was raised from 500 tons (for one month sailing) to 700 tons.

54. *Kaitensen hyakunen no ayumi*, 304–330. After the war, Shinohara Noboru regretted not having obtained a patent on his idea of inserting amplifiers inside the submarine cable. Such a technology was used by the United States in Key West–Havana cable as well as the transatlantic telephone cable. Shinohara, "Soka keburu kara musoka keburu e," *Teishin shiwa* 2:245; also in Nihon denshin denwa kôsha gijutsukyoku, *Denki tsûshin jiritshu gijutsu kaihatsu shi: Hansô denwa hen* (A history of the independent development of telecommunications technology: Carrier telephone) (Tokyo, 1972), 261–263, 42, 101–102.

55. Nihon Taiyô kaitei densen kabushiki kaisha, *Nihon Taiyô Kaitei Densen Kabushikikaisha Shi* (A history of the Japan Oceanic Submarine Cable Co.) (Tokyo, 1970), 72–73.

56. On raw material in Southeast Asia needed for telecommunications manufacturing, see "Denki kizai no shigensaki o kataru," and Arai Hiroshi, "Denki kizai no jûyô fuzoku shizai no taisaku," *Denki tsûshin* 5, no. 21 (8/1942): 40–61, 61–67. On gutta-percha, see Shimabara Sadakichi, "Marai no gutta percha ni tsuie," *Denki tsûshin* 6, no. 28 (9/1943): 21–28.

57. Tokyo kokusai tsûshin shinkôkai, *Ts'shin ni kansuru gun-kan-min kondaikai kiroku* (Minutes of a panel discussion concerning communications among the military, government and private sector) (June 1943).

58. Headrick, *Tools of Empire*, 12. The Taiwan cable, to be sure, would belong to the category of consolidation.

59. Peter Duus, "Japan's Wartime Empire: Problems and Issues," in *The Japanese Wartime Empire, 1931–1945*, ed. Peter Duus, Ramon Myers, and Mark Peattie (Princeton: Princeton University Press, 1996), xiv.

Conclusions and Commentary

12

The Geopolitical Implications of Communication under the Seas

Peter J. Hugill

Communication under the seas via submarine cables has taken three forms: submarine telegraph cables, submarine telephone cables, and submarine fiber-optic cables. The papers in this volume address all three in various ways, as well as land-based and wireless alternatives. The sustained activity surrounding submarine telecommunication defines a technological era of considerable importance to the development of a truly global economy—one that crosses the boundaries of periods that historians tend to make discrete. For example, Eric Hobsbawm's persuasive macrohistories of the era divide it into three periods: the "age of capital" (1848–1875); the "age of empire" (1875–1914); and the "age of extremes" (1914–1991).[1] The first of Hobsbawm's "ages" is almost coterminous with the start of both land-based and submarine telegraphy; the second is almost totally dominated by submarine telegraphy, although it saw the beginnings of wireless; and the last extends well beyond it, almost to the present, thus including submarine telephone cables, fiber-optic cables, and several wireless alternatives. Such division removes the essential unity of telecommunications effort that underpinned the emergence of the global economy under British leadership during the late nineteenth century, its temporary retreat in the first half of the twentieth century, and its full-blown return under American leadership in the second half. A similar problem exists at the level of the history of technology, with most works focusing on only one part of the technological picture, such as telegraphy,[2] wireless,[3] telephony,[4] satellites,[5] or fiber optics.[6] However good each of these works, none gives a complete picture.

Headrick's seminal work on the role of the submarine telegraph in European imperialism, in effect a continuation and expansion of a major theme in his earlier work on the technologies undergirding imperialism, suffers precisely this problem of historical compartmentalization.[7] From a somewhat different perspective, and within the three macrohistorical periods defined by Hobsbawm on the basis of economic and political events, the period 1858–1988 was clearly one macrotechnological era, albeit with significant changes to the efficiency of cables over time and the technologies of their construction and operation. The differential impact of those economic and political events on submarine telecommunication and the effects of submarine telecommunication on those economic and political events cannot be ignored.

Several works published since the conference address some of these interactions. Nickles addresses the relationship between submarine telegraphy and diplomacy, though within an almost entirely American and Atlanticist context. Winkler addresses both submarine telegraphy and wireless in his study of America's realization that its geostrategically important telecommunications depended heavily on Britain during World War I, again an Atlanticist perspective. Wireless seemed to offer a way around that dependence, although RCA would be quickly seduced from the U.S. Navy by the promise of vast profits from broadcast entertainment. More comprehensively, Winseck and Pike examine the role of both cables and wireless in the processes of imperialism and globalization between 1860 and 1930, with particular emphasis on Latin America and Asia. Each work suffers various flaws, mostly from a failure to blend analysis of the complex, emerging telecommunications technologies with their insights into their use in the diplomatic and economic contexts.[8]

By far the most important submarine telecommunication route has always been that across the Atlantic from northern Europe, primarily Britain, to North America, primarily the United States. Between 1858 and the present, three types of submarine cables have been laid. First came submarine telegraph cables, the first of which was installed in 1858. Being essentially an empirical venture, executed without a theoretical understanding of cable workings, this project soon failed. Success came in 1866, with better cable, better laying systems, and much better (though still imperfect) understanding of the electronics of transmission.[9] Each of these elements was then continuously improved. Message speed and capacity rose and cost fell. The last telegraph cable across the Atlantic was installed in 1928. As Finn notes in this volume, however, the very high capital cost of submarine cables tended to produce technological stagnation and open the field to technological challenges, a topic taken up here in Winkler's paper. The second type of cable was the submarine telephone cable, the first of which, Trans-Atlantic Telephone 1 (TAT-1), was installed in 1956, the last, TAT-7, in 1983. Third came the fiber-optic cables. TAT-8 opened for business in 1988.

Three "kissing cousins" developed alongside the three cable systems. First, low-frequency, longwave wireless emerged between Marconi's 1901 pioneering transmission across the Atlantic and the mid-1920s. Second came high-frequency, shortwave wireless, from the mid-1920s until the success of

TAT-1. These two eras overlapped the era of submarine telegraph cables and, despite some limited use for telephony, both were dominated by telegraphic messages. Third, beginning with the formation of INTELSAT in 1964, came the era of satellites, which has been brought to an end by the success of fiber optics.

Each of these telecommunications networks, three wired and three unwired, has been an explicitly geopolitical construct, integral to the process of hegemonic maintenance, challenge, and response that has been ongoing among the polities at the core of the capitalist world economy throughout the period. For this era, this process centered first on the maintenance and decline of British hegemony, then on the rise of America and imperial Germany as challenger powers, and finally on the success and maintenance of American hegemony.

This central point is easily missed in the tumultuous events that followed the failed hegemonic transition of the period immediately after the first German war. The position I argue from here is that of the world system theory developed in the work of Immanuel Wallerstein,[10] according to which the consequence of the failure of America to assume hegemony in the early 1920s was a two-stage war of hegemonic transition, best described as the German wars, the first extending from 1914 to 1918 and the second from 1939 to 1945, with an armistice dividing them. After Germany renewed its hegemonic challenge, a new challenger—Japan—entered the arena in 1941, requiring America to fight an expensive and protracted two-front war of transition. As Yang notes in this volume, submarine cables and wireless played a major part in the buildup of that Japanese challenge. In all these events, as I have argued in *Global Communications since 1844*,[11] telecommunications played a central role. Hegemonic shifts are themselves primarily driven by technological shifts, and in this case the dominant technology was telecommunications.

Technology itself is not neutral. Technology is a powerful social construct, and those who develop and use it have defined goals, even if they have not clearly stated or understood those goals. I also take the position that all technologies are not equal. Like the animals on Orwell's farm, some are more equal than others. The point is well stated by Peter Hall and Paschal Preston, who comment on the importance of the decline of old technology and the emergence of new technology in driving investment cycles and thus the periodic waves in the world economy first empirically observed by Nikolai Kondratiev.[12] Some technologies are simply more significant in the long run than others, and for such a technology Hall and Preston coined the term *carrier wave*. As they indicate, telegraphy—submarine cable telegraphy in particular—was very much part of the carrier wave that drove the emergence of the new information technology in the 1840s and 1850s. Starting on that basis, and accelerating very hard once Heinrich Hertz had begun a proper theoretical understanding of the physics of electromagnetic phenomena on the basis of James Clerk Maxwell's work,[13] a whole raft of information technologies developed through to the present. The consequence is that the physical universe we currently inhabit is defined as much by our understanding and application of Maxwell's equations to telecommunications as by Newtonian physics.

Carrier waves thus produce technologies that are focused on new ways of making or doing things or on much-refined understandings about how things can be made or done. The polity that best understands and implements such breakthroughs will improve its performance potential in the hegemonic struggle. At a global level, for the past hundred years or so we have come to understand that the struggle works itself out as an explicitly geopolitical process, with new state principles and technologies contributing to that process. The technologies and their development are often very simple from a scientific and engineering standpoint. The way they work out in the geopolitical realm is usually anything but.

For the sake of clarity I thus address four interconnected points around which the technology of submarine cables and telecommunications in general has swirled:

1. The emergence in the late 1800s of the new state principle of the territorially bounded nation-state—highly bureaucratized, increasingly militarized, with autarkic tendencies replacing liberal globalizing ones.
2. The emergence of discursive literacy and the rise of a broad middle class supported by and supportive of the nation-state.
3. The emergence of two geopolitical discourses in the period between 1880 and 1904: that of Alfred Thayer Mahan, with regard to the primacy of sea power, and that of Halford Mackinder, on the creation of continental states and their likely future dominance over the maritime powers.
4. The structure of hegemonic transitions in the world economy and the question of how the world economy is structured and how transitions are driven.

From Liberalism toward Autarky and Back Again: The Role of the Nation-State

Inasmuch as cables emerged in the 1840s over land and in the 1850s under the seas, they were born into a world dominated by liberal globalism. From Britain's repeal of its mercantilist Corn Laws and Navigation Acts in the late

1840s until the return to tariff protection by the United States in 1861, the world economy was ruled by free trade and the ideals of Britain's Manchester liberals. In this context, Robert Boyce's paper in this volume is absolutely on target. The City of London was obsessed with securing its place at the center of the global capitalist enterprise. It achieved this from the late 1860s on by financing a global network of submarine telegraph cables that centered on London. George Parkin would call the submarine cable network the "nerve system" of the empire.[14] In truth, however, it began as the nerve system of the City of London at the height of liberal globalism and evolved into the geopolitical construct described by Parkin.

Under liberal globalism Britain was able to acquire needed commodities—raw materials for its industries and food for its burgeoning urban industrial populations—by the sale of the goods produced by those industries and, increasingly, by the sale of the financial services of the City of London. After 1861 the imposition of first American, then German, tariffs increasingly restricted the sale of British manufactures. Britain's industrial performance faltered. In part this was because of the failure to restructure British industries to meet the needs of the second industrial revolution just getting under way; in part it was because returns to British capital on overseas investment were higher than returns to capital on British industrial restructuring. British financial services thus had to expand their market in order to ensure the exchange required for purchase of raw materials and food. Intensification of the sale of goods and services required even more telecommunications than had been needed merely for the flows of commodities under liberal globalism.

The increasing failure of liberal globalism after 1861 was associated with two new and increasingly dominant themes in global politics: the triumph of the territorially bounded nation-state as the dominant state principle and the "new imperialism." The French had pioneered the idea of the territorially bounded nation-state as part of the French Revolution, but their failure to wrest hegemony from a rapidly industrializing Britain in the Napoleonic Wars had allowed Britain to impose liberal globalism on the world economy of the first two-thirds of the nineteenth century. At the end of the Napoleonic Wars Britain was the world's only manufacturing nation; it also possessed the world's most powerful navy. By 1815 British hegemony was thus absolute in the three arenas defined in Wallerstein's 1984 model: agro-industrial production, commerce, and finance. As Wallerstein points out, free trade and globalization—that is, liberal globalism—are the logical policies of both the hegemonic core power and of peripheral polities in the world economy that trade their agricultural raw materials and food for the industrial products of the hegemonic core state.

Two events moved the world toward adoption of the French concept of the territorially bounded nation-state. First, the concept was manifested by the Union at the outbreak of the American Civil War. Despite the Founders' claims that American sovereignty was vested in a people rather than in a territory, the Civil War was fought in part to ensure the territorial integrity of the American state, however much one part of that polity wished to secede as an independent people. In 1861 America polarized into two polities. As a peripheral—in Wallerstein's terms—producer of a vital industrial raw material, cotton, the South went so far as to forbid, in Section 8 of the Constitution of the Confederate States of America, "duties and taxes on importations . . . to promote or foster any branch of industry."[15] The Union adopted both the broad framework of a territorially bounded nation-state on the French model and the economic policy of protectionism. In 1865, when the Northern polity emerged triumphant, it was able to impose its economic will on the defeated Confederacy.

The second event was the implicitly irredentist claim of Bismarck to unify at least most of the German-speaking peoples in a defined German territory upon unification in 1871, a policy later made explicit and pushed to its limits by Hitler.

Both America and Germany moved quickly to impose semi-autarkic views of the economy on the territorially bounded states within their writ, rejecting the liberal globalism of Britain and thus reducing Britain's access to commodities and markets. As Wallerstein noted, protectionism of some form is the logical and preferred economic policy of hegemonic challengers in a period of multipolarity.[16] The United States embarked on a series of ever more substantial tariffs in 1861, upon the secession of the Confederate States, tariffs not seriously reduced until the Bretton Woods agreement of 1944. Bismarck's Germany followed a like path in 1879, to deal with macroeconomic imbalances driven by the Franco-Prussian War and the stresses of unification.[17]

The industrial structure of the world economy of the late 1800s was still heavily biased to organic inputs, especially in the textile industry, then still the leading sector of most industrialized economies. Food was needed for the populations of the burgeoning industrial cities, both human and animal, and by the late 1800s countries like Britain were importing huge amounts of food. Even where population pressure and urbanization were less significant, such as in France, or where heroic efforts were being made to improve the output of domestic agriculture, as in Germany, European agriculture was environmentally incapable of providing tropical or subtropical goods: foods such as sugar; raw materials for industry and telecommunications, such as cotton, silk, rubber, and gutta-percha; the stimulants to which industrial populations had become addicted, such as cacao, coffee, and tea; the drugs Europeans needed to survive in the tropics, such as cinchona for antimalarial quinine; and the drugs they needed to kill pain or feed addiction, such as opium. One huge consequence of the emergence of the territorially bounded nation-state was thus an ever-growing concern, especially on the part of Germany, with access to tropical and subtropical staples, which, with the failure of liberal globalism, required the possession of appropriate chunks of territory as colonies. Even America, generally blessed with the subtropical regions of the failed Confederacy, entered such a phase in its imperial adventures in Cuba and the Philippines. We now refer to this period of late 1800s behavior by core or aspiring core powers as the period of "new imperialism."

Unmistakable signs of the new imperialism first clearly emerged in the behavior of the French after their loss of the Franco-Prussian War of 1871. Colonies, cheaply acquired and cheaply run, rebuilt French prestige as compensation for the humiliating loss of that war.[18] Such behavior required secure communications between the imperial center and the colonial outpost, and the French jury-rigged at least an attempt at a global telecommunications system to counter Britain's. By the 1880s, however, French behavior was a sideshow in an increasingly confused period of multipolarity caused by the decline of British hegemony resulting from the increasing failure of liberal globalism.

The two polities that vied to succeed Britain, America and Germany, began in the late 1800s to embrace both an aggressive nationalism and, in the aftermath of the Berlin Conference of 1884, the new imperialism. Driven by a desire for the prestige that colonies were then seen as conferring, Germany hosted the Berlin Conference in part to secure for itself sources of tropical and subtropical goods. By 1884 about the only space on the map left uncolonized by such older powers as Britain and France was Africa—hence the conference's sobriquet, the "Grab for Africa."

As both Germany and America strove for an increasingly imperial and global presence, both also felt the need to replicate Britain's global telecommunications system. Britain's telecommunications system was predicated upon the ideals of liberal globalism. British cable companies willingly carried the messages of any who were able to pay. But, in the increasingly paranoid political environment brought about by the combination of the new nation-state structure and the new imperialism, each polity came to distrust its perceived opponents. Submarine telegraphy became a strategic and military issue, as Headrick's paper in this volume so clearly shows. Ultimately all telecommunications became a geopolitical issue, first for the French, then for the Germans and Americans, and lastly for the British themselves. French and German attempts to replicate Britain's submarine cable network both failed. What made the British network so effective was its origin in the era of liberal globalism. Most of Britain's cables made a profit because they had been laid to make a profit. French and German cables never did because they were laid to serve geopolitical needs. Even so, by the late nineteenth century Britain fell into the same trap of laying geopolitically motivated submarine cables. When the British realized that this effort was unlikely ever to prove profitable, they embraced the emergent technology of wireless. Despite failure to dominate the first, rather expensive wireless technology, Britain quickly came to dominate the second, very much cheaper wireless technology. The American geopolitical challenge emerged in the first wireless technology and failed. U.S. success came much later, with the second submarine cable technology (telephone cables) and the third wireless technology (telecommunication satellites). The switch to the third submarine cable period, that of fiber-optic cables, leveled the playing field once again, leaving America and Europe relatively even.

American and German tariffs, plus the failure of Britain to embrace the new, science-based industrial technologies of the second industrial revolution, caused increasing problems in Britain's manufacturing sector, leading many to conclude that Britain too needed to embrace protectionism. Some politicians, such as Joseph Chamberlain, who represented Birmingham, a manufacturing city hit heavily by British industrial failures, saw protectionism as a way of rescuing the working class from otherwise inevitable deindustrialization. Others, such as British economist J. A. Hobson[19] and geopolitician Halford Mackinder,[20] saw a renewed empire as a secure market for British goods after the relative closure of the American and German export markets by high tariffs. Both Chamberlain and Mackinder left the Liberal Party for the newly formed imperial protectionist wing of the Conservative Party, helping push Britain into even more firmly embracing the new imperialism. Britain's relative failure in the second industrial revolution also helped persuade British investors that foreign and financial investments were more secure than industrial ones. The rapidly increasing efficiency of American and German factories in, for example, such traditional British export fields as cotton textiles, meant that Britain's competitors could pay more for raw cotton than much less efficient British factories. As a consequence the British began, from the early 1900s on, to seek geopolitically secure sources of cotton within the boundaries of the empire rather than rely on the United States.

Again, this focus on investment overseas emphasized the need for global telecommunications, and the British began to install subsidized cables to parts of the empire that lacked enough commodity trade to make cables profitable.

The new technology of wireless, offered first to Britain by Guglielmo Marconi, promised cheaper fixed costs than submarine cables. Marconi's career—beginning at the close of the era of British liberal globalism and ending with the autarkic nationalism of fascism—reflects both the political and technological issues discussed here. Born of an Italian father and an Irish mother, from the wealthy Jameson family of Irish whiskey fame, Marconi came to Britain in 1896 at his mother's behest, she having failed to interest the Italian authorities in his breakthrough.[21] When the British General Post Office refused to offer a decent price for his technology, Marconi used his mother's connections in the distilling industry to capitalize the Wireless Telegraph and Signal Company in London in 1897 and installed his cousin Henry Jameson-Davis as managing director.[22] Marconi's technology saw its first effective use in that epitome of liberal globalism, Britain's merchant fleet, driven by Lloyd's refusal to give its highest insurance rating to a ship unless it was equipped with Marconi wireless. The wisdom of this policy was driven home dramatically by the *Titanic* disaster of 1912.

In 1911 the British state, increasingly attracted to the new imperialism, embraced Marconi's proposal for a chain of low-frequency, longwave wireless telegraph stations—the Imperial Wireless Chain—to supplement the submarine cables. The first German war forced a great deal of telecommunications technology, and by the mid-1920s high-frequency, shortwave Marconi beam stations were delivering cheap and reliable imperial telecommunications. Marconi himself, upset at the forced merger of his company with the submarine cable interests to form Imperial and International Communications in 1929 (renamed Cable and Wireless in 1934), embraced fascism and returned to Italy in the early 1930s.

The slightly later Danish failure to embrace the superior, though still low-frequency, wireless technology developed by Danish inventor Poulsen and mentioned here in Jacobsen's paper is as instructive as the shortsightedness of the Italian Ministry of Posts and Telegraphs with regard to Marconi's technology. For certain types of technological innovation geopolitical context is a central part of the adoption environment. However great the success of the Danish Great Northern Telegraph Company in installing landlines across the Russian Empire from Europe to Asia, they were unlikely to see the advantage of wireless because they had no maritime, global context into which to place it. Much of the reason for the success of wireless was that it increased the safety of a nation's merchant ships and that it gave a nation's navy centralized command and control of ships at sea.

DISCURSIVE LITERACY, THE NATION-STATE, AND THE RISE OF THE MIDDLE CLASS

My second context grows, in part, out of the first and is based on what Max Weber observed at the time: the emergence of a fundamentally new type of nation-state in late-nineteenth-century Germany. As Michael Mann has noted,[23] the idea of the territorially bounded nation-state had two consequences that led to the bureaucratization and militarization of such states. Under liberal globalism the British state was no more than a "night watchman" state, policing the seas of the world to ensure the free movement of commodities and manufactured goods. Managing the new, nation-state economies, however, became a job for government. Increasingly complex economies could no longer be left solely to market forces. In polities with a strong liberal tradition, such management appeared partly as regulation of the economy by the nation-state and partly as more careful management of production within firms to avoid excessive regulation. Polities with a more centralized governance tradition managed the economy more directly, up to and including government ownership of the means of production. All such management required bureaucratization, whether at the level of the nation-state or the firm, which in turn required a much more educated workforce. All such management required much more in the way of telecommunications.

Creating a more educated workforce thus became a central effort for all nation-states at the core of the world economy. Compulsory secondary education was deemed advantageous to the nation-state after the military lessons of the U.S. Civil War and the Prussian wars of the 1860s and 1870s. This caused a rapid increase in the size of the middle classes and an equally rapid rise of discursive literacy among the working classes in the 1880s and 1890s. This in turn accelerated demand for reading material in at least four forms: popular daily newspapers, often called the "yellow press" because such papers were printed on pulp paper that aged rapidly; weekly, illustrated news or opinion magazines; monthly, illustrated magazines; and cheap or "dime" novels. The first three of these, but especially the first, markedly increased demand for global telegraph services, which had always had such news services as Reuters and Associated Press as their major customers; nationalism and imperialism did the rest.

Popular newspapers found that whipping up national fervor, especially over imperial issues—even to the point of fomenting wars, as William

Randolph Hearst would demonstrate over the sinking of the USS *Maine*—sold huge numbers of daily papers. Although the popular press was by far the biggest market for global news spread via submarine cables, the weekly and monthly magazines contributed their share. By 1900 all the core and aspiring core powers in the world economy of the time had imperial interests. Many weekly and opinion magazines dealt with imperial issues. To take but one example of many, the German illustrated weekly magazine *Simplicissimus* paid much attention to German colonial events, portraying the leaders of other nations as corrupt and decadent—an inebriated Queen Victoria is shown plucking South African ostriches during the Boer War; American President Teddy Roosevelt is shown as an uncouth, gun-toting cowboy. Such news had to be as current as that of the daily papers. *National Geographic*, published by America's National Geographic Society, was transformed from an academic journal to a highly profitable mass circulation monthly magazine on the basis of its reporting of American acquisition of the collapsing Spanish Empire, as Susan Schulten has shown.[24] This lesson was not lost on other publishers. Colonial affairs sold and colonial affairs required good telecommunications.

The bureaucratic nation-state needed a large, relatively well-educated labor force. This emerged as a broad middle class of teachers, professors, police, and civil administrators, layered on top of the traditional state functionaries in the militaries, which were also gaining rapidly in numbers of personnel in the late 1800s. Imperialism increased demand both for navies, as much for prestige as for security, and for armies to subdue revolts in the colonies. An expanded middle class, dependent on the nation-state for its livelihood, strongly embraced nationalism. The more people worked for the nation-state or, post-Bismarck, received transfer payments from it in the form of educational grants, compensation for injuries, retirement benefits, and health care benefits, the more people identified with the nation-state. People paid taxes to the nation-state and received a package of real or promised services in return. This set up an incipient tension between the liberal globalism of the first half of the nineteenth century, exemplified in the British "night watchman" ideal of the state, and the nationalism of the second half of the century, developed most strongly in such nation-states as the United States and Germany. Britain initially resisted nationalism because of the power of the City of London to enforce the ideals of liberal globalism and because the strength of Britain's global telecommunications network continued to promise British capital a way of continuing to make profits without embracing protectionism.

We rightly see the first German war as marking the death of liberal globalism—hence Hobsbawm's claim that it marked our entry into the age of extremes. Liberal globalism, however, returned at the Bretton Woods agreement of 1944 and was secured by American hegemonic victory at the close of the second German war in 1945. It is impossible here not to comment on the implications of U.S. domestic politics for telecommunications. The Republican Party, which had cemented its power by the Morrill tariff of 1861, was a classic cadre party, defending the special interests of those it saw as maintaining it in power—the industrialists who profited from the protectionist policies of the territorially bounded nation-state. The Republican Party's failure to cope with the Depression of 1929 allowed the Democrats, just making the turn to being America's first mass party, to return to power and to begin to restore liberal globalism, which they moved a long way toward at the Bretton Woods agreement of 1944.

Cadre and mass parties clearly see different needs for telecommunications. As a cadre party committed to a nation-state perspective and the protection of American industry, the Republican Party of the late 1800s and early 1900s, along with their fellow travelers in the protectionist wing of the Democratic Party, saw U.S. telecommunications as merely about national security, hence their general support of the U.S. Navy's attempt to control wireless telecommunications via the creation of the Radio Corporation of America (RCA). As a mass party committed to an internationalist perspective, the post-Depression Democrats saw the need to radically expand the global reach of America's telecommunications to capture a larger share of world trade. By the 1950s the obvious way to achieve this was by encouraging Bell's burgeoning system of submarine telephone cables. The older, national security perspective was, however, scarcely lost. Driven by Cold War fears, even Democrats could be realists, and the national security perspective continued side by side with renewed liberal globalism after the German wars were over. What RCA had signally failed to accomplish in the 1920s was, effectively, achieved for national security through the formation of INTELSAT in the 1960s.

The Geopolitical Models of Mahan and Mackinder

Between the publication in Germany of Friedrich Ratzel's ideas about the organic nature of the state (another new state principle) and the publication in Britain in 1904 of Halford Mackinder's "Pivot" paper, the theory behind which was later renamed the "heartland thesis," the idea of geopolitics emerged. A critical element in that emergence, downplayed in the writings of nearly all commentators on geopolitics except Mackinder himself, was the work of U.S. admiral Alfred Thayer Mahan. Mahan's definitive works, *The*

Influence of Seapower on History (1890) and *The Influence of Seapower upon the French Revolution and Empire*,[25] drew on Dutch, British, and French history to argue that sea power was central to achieving or not achieving global power. Mahan's work thus could be, and was, read by newer powers as a recipe for achieving global power. Notable adherents of Mahanian reasoning included Teddy Roosevelt, who gave the books a glowing review in *Atlantic Monthly*,[26] and Kaiser Wilhelm II. The kaiser noted that he was "trying to learn it by heart. . . . It is aboard all my ships and constantly quoted by my Captains and officers."[27]

Inasmuch as we can derive a geopolitics from Mahan, it is that the history of the European expansions from the 1400s on was driven by sea power. Maritime states that derived their income from trade displaced in importance territorial states that derived their income from land. The best known writing on this, and on the existence of maritime and territorial polities within the same state, comes from the work on French political history by Edward Whiting Fox, although Fox's thesis—that there are "two Frances," one a maritime polity and one a territorial polity—was actually stated earlier by Mackinder for Germany. In France's case the maritime and territorial polities struggled for control of the Revolution—a struggle that the maritime polity lost, leaving the territorial polity to control France's destiny.[28] The consequence was that the French came to conceive "of naval warfare as merely a branch of land warfare" and were thus never able to challenge Britain for command of the sea.[29]

When a territorial state takes on a trading state, it loses, as Mahan pointed out in his analysis of the failure of France, Napoleon's territorial strategy failing dismally against Britain's maritime one. Mahan exemplified British commercial and military policies of the late 1700s in the work of two great prime ministers, William Pitt the Elder and William Pitt the Younger, whose policies were

> to extend and consolidate . . . [Britain's] sea power; to lay the world under contribution to her commerce; to control the sea by an all-powerful navy; to extend her colonial empire by conquest, thereby increasing her resources, multiplying her naval bases, and depriving her enemy alike of revenues and points whence he could trouble English shipping; to embarrass the great enemy, France, by subsidizing continental allies . . . [subsidies] only the sustained commercial prosperity of England could supply.[30]

The consequences of British maritime, trade-based policies for the French territorial state, a state codified by Napoleon in his Continental System, were, ultimately, ruinous.

> By the mastery of the sea, by the destruction of the French colonial system and commerce, by her persistent enmity to the spirit of aggression which was incarnate in the French Revolution and personified in Napoleon, by her own sustained and unshaken strength, [Britain] . . . drove the enemy into the battle-field of the Continental System, where his final ruin was certain.[31]

Germany did no better, in either of its wars. Mackinder analyzed German entry into the first German war as a struggle between "Hamburg and overseas dominion, or Bagdad and the Heartland": the consequence was that "Berlin . . . fought on two fronts without making up her mind on which front she meant to win . . . her strategical aim was . . . uncertain."[32] In the second German war Hitler repeated the most disastrous of Napoleon's mistakes in his desire to expand territorially.

The term *Geopolitik* was not coined by Ratzel (who referred merely to "political geography"), Mahan, or Mackinder; rather, it was the invention of Swedish geographer Rudolf Kjellén, in 1899.[33] The term was popularized by German geographer Karl Haushofer. After the German loss of World War I, Haushofer, through his student Rudolf Hess, became embroiled with Hitler during his jail term in Munich. Haushofer's Institut für Geopolitik at the University of Munich became a malignant force in German geography and culture in the Nazi period.[34] German *Geopolitik* was about hegemonic ambitions and expansion: British geopolitics was a way of trying to understand and cope with what seemed likely to be hegemonic decline. Notably, Mackinder's model of the possible causes of such decline was driven specifically by the emergence of the new technologies—most notably the railroad and the telegraph—creating the bureaucratic, continental, national states. Mackinder's prescription for avoiding decline was to abandon the liberal globalism evident in his earliest publications (along with his membership in the Liberal Party)[35] and embrace the new imperialism by joining the imperial protectionist wing of the Conservative Party. As he put it in 1902 in *Britain and the British Seas*,

> Under a condition of universal free trade, the dream of the sixties of the last century, industrial life and empire might be dissociated, but when competing countries seek to monopolise markets by means of customs tariffs, even democracies are compelled to annex empires. . . . It is difficult to see how foreign governments . . . could have resisted the pressure put upon them for protection against a trading nation which had insular security and a long lead in the race. Under a condition of universal free trade such a nation would almost inevitably increase its lead, and might

> ultimately reduce the whole world to economic subjection. Hence the most important facts of contemporary political geography are the extent of the red patches of British dominion upon the map of the world, and the position of the hostile customs frontiers.[36]

Mackinder's main fear, expressed most cogently in his 1904 article "The Geographical Pivot of History," was that railroads could be used to expand out of the pivot area to unite the Eurasian landmass, the "World-Island," under the control of one polity. Such a polity would be both beyond the abilities of the maritime polities to project their power inland and possessed of such huge resources of energy and raw materials that it would be able to build the world's greatest fleet: "The empire of the world would then be in sight."[37] After the first German war, in *Democratic Ideals and Reality*, Mackinder renamed his pivot area the "Heartland" and elaborated his argument to include telegraphs and air transport as tools of continental integration, concluding

> Who rules East Europe commands the Heartland:
> Who rules the Heartland commands the World-Island
> Who rules the World-Island commands the World.[38]

Trading states, maritime in nature for the past five hundred years, have a particular interest in long-distance communications, for their ability to make wealth depends on their ability to access information through which they can control trade. It was no accident, in geopolitical terms, that Britain, the archetypal trading state, developed the world's first effective global telecommunications system based on submarine cables, and it has been no accident that Britain has continued to invest heavily in telecommunications. Equally, the relative failure of France and Germany was, in part, a geopolitical failure. Since both had dominant territorial polities within the boundaries of their nation-states, they failed to focus adequate resources on the needs of their much less powerful trading polities, and thus they were unable to tap into the trade flows that enriched Britain.

The U.S. case, however, is unique. Before 1776 America was part of the world's greatest trading state, Britain. After 1783 America developed as an archetypal territorially bounded nation-state, albeit retaining strong trading components around its Atlantic and Gulf coasts. However, America's territorial competitors were easily overcome (Mexico and the indigenous Indians) or too strong (British Canada), so that America did not waste huge amounts of resources during its phase of territorial expansion. Although two polities developed in America during the Civil War era, the split was not along trading/territorial lines. Both Northern and Southern polities had trading as well as territorial components.

Once America's territorial expansion was complete, it returned, albeit slowly, to its roots as a trading state. I argue that that history helps, in part, to explain how U.S. telecommunications developed. In its territorial expansion phase, the United States had no real interest in telecommunications beyond the boundaries of the nation-state: such international services as were needed could be bought from British providers, who were, in any case, the main providers of capital in the American market. The first American attempt to develop a global network, using low-frequency wireless, was grounded in the geopolitical assumptions of a territorial state—that only national security mattered. The attempt to dominate low-frequency wireless was thus more than likely to fail, since the only really interested party was the navy, and the technology was driven by the interests of the navy alone.

The reassertion of the geopolitical concerns of a trading state after the second German war resulted in the successful installation of the second submarine cable system by Bell Telephone, a classic commercial network with no overt geopolitical agenda. The national security demands of the Cold War caused a reassertion of the overt geopolitical agenda inherent in the development of INTELSAT, which proved a commercial success because its fixed costs were relatively low. That in turn gave way to a further elaboration of the telecommunications interests of the trading state, the fiber-optic cable network. U.S. national security obsessions about the Cold War combined with the success of INTELSAT to blind Americans just enough to the emergence of fiber optics that they were unable to achieve total dominance in the technology. As Jeff Hecht's paper in this volume notes, much of the technology that went into fiber optics was envisioned in Britain by Alec Reeves before being developed as a practical technology in Britain and the United States. Fiber optics indeed was guilty of "turning the tables on satellites."[39] Its unintended consequence was to end American telecommunications hegemony.

Structure and Causation of Hegemonic Transitions

Although arguments persist in the world system literature about what drives hegemonic transitions, some authors, myself included, follow the arguments of Kondratiev and Joseph Alois Schumpeter that there are regular upswings and downswings in the world economy. These waves last approximately 55 years and seem driven by technological change. George Modelski suggested that two Kondratiev waves make up a world leadership cycle, of which we have seen five since the mid-1400s. In 1988 Hall and Preston suggested that some technologies have been far more important than others in driving Kondratiev upswings. They identified telecommunications and information technologies as important in the last three upswings, beginning respectively in the 1840s, late 1890s, and mid-1900s. Telegraphy, first terrestrial then submarine, drove the first. Wireless, first for telecommunications then for entertainment, drove the second, abetted by recorded music and sound. A mélange of entertainment devices, prominent among which was television, with its associated information technology of radar, drove the last. When *The Carrier Wave* was published, Hall and Preston were forecasting that personal computers and the Internet would drive the Kondratiev upswing now under way. The fact that this Kondratiev upswing coincides, by Modelski's timetable and mine, with the likely start of the next world leadership cycle makes this question timely.

As a social scientist, I recognize that historians are leery of prediction, but I also believe that history is too important to be left to historians alone. Prediction *is* important, not least because, as Goldstein argues, long cycle shifts are intimately linked to cycles of war, with wars of transition being the norm at hegemonic junctures.[40] These wars are almost always military ones, although economic war is also part of military conflict and it is possible that a transition war could be a purely economic one. The last two such junctures saw bloody global wars of transition: the Napoleonic Wars, which ushered in Britain's nineteenth-century hegemony, and the German wars, which ushered in America's twentieth-century hegemony.

So let me suggest three predictive possibilities, the accuracy of which time will tell. They are, quite frankly, in my order of preference:

1. This time liberal globalism does not fail. Open (or relatively open) diffusion of science and technology ensures that no nation can develop any marked advantage over any other and, in a world linked by high-speed telecommunications, a global community emerges. We exit the cycle of hegemonic competition.
2. The hegemonic cycle recurs, with or without a war of transition. Having bet on the right technologies and the necessary restructuring of the economy and society to support them, as can be inferred from the sustained economic boom of the 1990s, American hegemony (currently cultural and military, but not economic) resumes. Given current American investment and political behavior, the right technologies would have to be in current advanced information technologies and the Internet, in such emerging technologies as nanotechnology, and in such predictable technologies as quantum computers.
3. As in the second possibility, but America fails to make the right technological bets and another hegemon emerges. The current likely contenders are the European Union and Japan, both of which generate gross national product

(GNP) per capita figures approaching those of America. However, and despite the recent falloff in American economic performance, both regions have obvious economic failings in that they have not adjusted their economies well to recent technological change and neither is strong militarily. China is a highly unlikely contender at present because of its economic weakness when measured in GNP per capita. In this model the right technological bets will likely be in such areas as genetic engineering, cloning, stem cell research, and environmental science, all of which have encountered strong ideological opposition recently in America and much less elsewhere.

Lest the last possibility sound unlikely, I argued in *Global Communications since 1844* that betting on the wrong carrier wave technology was precisely one of the things that delayed U.S. hegemony from the end of the first to the end of the second German war. In that book I argued that the U.S. Navy preferred to stay with low-frequency wireless because of its demonstrated range. This failure to embrace high-frequency wireless was a classic failure of the geopolitical imagination, with deep roots in the U.S. society of its time. The dominant Republican Party was, in the 1920s, still a classic cadre party, beholden to industrialists and committed to the ideals of protectionism to maintain industrial profits within the territorially bounded nation-state. What might have been a global leavening of imperialism, as it was to the Republican Party of Teddy Roosevelt's era, was depressed by the corrosive domestic politics of isolationism, a politics fed and watered by Washington's warnings against foreign entanglements, the Monroe Doctrine, and the all-too-vivid images provided by the first German war of what came of ignoring the wisdom of the Founders. The navy's obsessions were in national security, which might have been fine had there also been a leavening of either imperial or commerce-based needs to protect U.S. interests abroad. As it was, the navy's chosen instrument in low-frequency wireless, RCA, was much less capable of picking up the imperial and commercial implications of shortwave, high-frequency wireless than was Marconi's in Britain. Marconi's, the British General Post Office and, later, Cable and Wireless, all fully understood the remarkable advantages offered by shortwave of low initial and running costs and ease of operation. Indeed, the fact that the British state forced the merger of the submarine cable interests, Marconi's, and the Post Office demonstrates the extent to which the government operated in the geopolitical realm. Shortwave wireless fitted exactly the needs of a far-flung global economic system in which many areas could not be affordably connected by high fixed-cost submarine cables. The United States' failure to move into shortwave wireless slowed its rise and Britain's decline, helped create a power vacuum in which imperial Japan emerged as a secondary hegemonic challenger to Germany, and, together with a host of other factors, helped ensure a second German war. A failure of imagination in a carrier-wave technology can carry a high geopolitical price.

Notes

1. Eric J. Hobsbawm, *The Age of Capital, 1848–1875* (London: Weidenfeld & Nicholson, 1975), *The Age of Empire, 1875–1914* (New York: Pantheon, 1987), *The Age of Extremes: A History of the World, 1914–1991* (New York: Pantheon, 1994).

2. Ken Beauchamp, *History of Telegraphy* (London: Institution of Electrical Engineers, 2001).

3. Hugh Barty-King, *Girdle Round the Earth: The Story of Cable and Wireless and Its Predecessors to Mark the Group's Jubilee, 1929–1979* (London: Heinemann, 1979).

4. E. F. O'Neill, *A History of Engineering and Science in the Bell System*, vol. 7, *Transmission Technology (1925–1975)* (Indianapolis: AT&T Bell Laboratories, 1985).

5. Heather E. Hudson, *Communication Satellites: Their Development and Impact* (New York: Free Press, 1990).

6. Jeff Hecht, *City of Light: The Story of Fiber Optics* (New York: Oxford University Press, 1999).

7. Daniel R. Headrick, *The Tentacles of Progress: Technology Transfer in the Age of Imperialism, 1850–1940* (New York: Oxford University Press, 1988); Daniel R. Headrick, *The Invisible Weapon: Telecommunications and International Politics, 1851–1945* (New York: Oxford University Press, 1991).

8. David Paull Nickles, *Under the Wire: How the Telegraph Changed Diplomacy* (Cambridge, Mass.: Harvard University Press, 2003); Jonathan Reed Winkler, *Nexus: Strategic Communications and American Security in World War I* (Cambridge, Mass.: Harvard University Press, 2008); Dwayne R. Winseck and Robert M. Pike, *Communication and Empire: Media, Markets, and Globalization, 1860–1930* (Durham: Duke University Press, 2007).

9. Vary T. Coates and Bernard Finn, *A Retrospective Technology Assessment: Submarine Telegraphy; The Transatlantic Cable of 1866* (San Francisco: San Francisco Press, 1979).

10. Immanuel Wallerstein, *The Modern World-System*, vol. 1, *Capitalist Agriculture and the Origins of the Modern World-Economy in the Sixteenth Century* (New York: Academic Press, 1974).

11. Peter J. Hugill, *Global Communications since 1844: Geopolitics and Technology* (Baltimore: Johns Hopkins University Press, 1999).

12. Peter Hall and Paschal Preston, *The Carrier Wave: New Information Technology and the Geography of Innovation, 1846–2000* (London: Unwin Hyman, 1988).

13. J. G. O'Hara and W. Pricah, *Hertz and the Maxwellians: A Study and Documentation of Electromagnetic Wave Radiation, 1873–1894* (London: Peter Peregrinus, 1987).

14. George R. Parkin, "The Geographical Unity of the British Empire," *Scottish Geographical Magazine* 10 (1984): 225–242.

15. *Protectionist* 10 (1909): 326.

16. Immanuel Wallerstein, "The Three Instances of Hegemony in the History of the Capitalist World-Economy," in *The Politics of the World-Economy*, ed. Immanuel Wallerstein (New York: Cambridge University Press, 1984), 37–46.

17. Asaf Zussman, "The Rise of German Protectionism in the 1870s: A Macroeconomic Perspective 2008" (pluto.huji.ac.il/~azussman/protectionism.pdf).

18. Headrick, *The Tentacles of Progress.*

19. J. A. Hobson, *Imperialism: A Study*, 3rd ed. (London: Allen & Unwin, 1938). First edition 1902.

20. Halford J. Mackinder, *Britain and the British Seas* (London: Heinemann, 1902).

21. Susan J. Douglas, *Inventing American Broadcasting, 1899–1922* (Baltimore: Johns Hopkins University Press, 1987), 17.

22 Ibid., 65.

23. Michael Mann, *The Sources of Social Power*, vol. 2, *The Rise of Classes and Nation-States, 1760–1914* (New York: Cambridge University Press, 1993).

24. Susan Schulten, *The Geographical Imagination in America, 1880–1950* (Chicago: University of Chicago Press, 2001).

25. Alfred Thayer Mahan, *The Influence of Sea Power on History, 1660–1783* (Boston: Little, Brown, 1890); *The Influence of Sea Power upon the French Revolution and Empire, 1793–1812*, 2 vols. (Boston: Little, Brown, 1892).

26. Theodore Roosevelt, *Atlantic Monthly* 66 (October 1890): 563–567; 71 (April 1893): 556–559.

27. Quoted in William E. Livezey, *Mahan on Sea Power*, 2nd ed. (Norman: University of Oklahoma Press, 1981), 67.

28. Edward Whiting Fox, *History in Geographic Perspective: The Other France* (New York: Norton, 1971), 13–15.

29. Margaret T. Sprout, "Mahan: Evangelist of Sea Power," in *Makers of Modern Strategy: Military Thought from Machiavelli to Hitler*, ed. Edward Meade Earle (Princeton: Princeton University Press, 1943), 433.

30. Mahan, *The Influence of Sea Power upon the French Revolution and Empire*, vol. 2, 389–390.

31. Ibid., 400–401.

32. Halford J. Mackinder, *Democratic Ideals and Reality: A Study in the Politics of Reconstruction* (1919; repr., New York: Henry Holt, 1942), 154.

33. Brian W. Blouet, *Geopolitics and Globalization in the Twentieth Century* (London: Reaktion, 2001), 30.

34. Derwent Whittlesey, "Haushofer: The Geopoliticians," in *Makers of Modern Strategy*, 397–407.

35. Halford J. Mackinder, "The Great Trade Routes (Their Connection with the Organization of Industry, Commerce, and Finance)," *Journal of the Institute of Bankers* 21 (1900): 1–6, 137–155, 266–273.

36. Mackinder, *Britain and the British Seas*, 342–343.

37. Halford J. Mackinder, "The Geographical Pivot of History," *Geographical Journal* 23 (1904): 436.

38. Mackinder, *Democratic Ideals and Reality*, 150.

39. Hecht, *City of Light*, 212–213.

40. Joshua S. Goldstein, *Long Cycles: Prosperity and War in the Modern Age* (New Haven: Yale University Press, 1988).

Appendix: Archival Sources for the History of Submarine Cables

Within the endnotes of the various contributions one can gain some feeling for the richness of the resources available for studying the history and impact of the cables. To make the search easier, we have assembled here, in narrative form, a survey of where some of the participants looked and what they found.

Sources in the United Kingdom and the United States (Jonathan Winkler)

United Kingdom

The history of Britain's role in global communications in the past two centuries is extensive, and any meaningful analysis must rest on both government and corporate records. The essential British government records are located at the National Archives (formerly the Public Record Office) in Kew Gardens, London. They include the Cabinet papers (CAB), the Admiralty records (ADM), and the Foreign Office records (FO), especially FO 371 and FO 368, relating to political and economic policy respectively. Treasury (T) and Board of Trade records (BT) are also useful. The General Post Office (GPO) operated the national telegraph and telephone system until its privatization at British Telecom, and government records relating to the telegraphic operations are now held either in the Post Office (Consignia) Archives, Freeling House, Mount Pleasant Complex, London, or at the BT Group Archives, Holborn Telephone Exchange, High Holborn, London. These are public records in private hands, and so it is best to contact the archives in advance of your research to ensure access. Cable and Wireless, Ltd., retained the collected corporate records of its various predecessors. These indispensable materials are located at the Porthcurno Telegraph Museum in Porthcurno, beyond Penzance, Cornwall. The absolute necessity of examining these records in any serious work on British international communications is offset by the absolute beauty of the archive's location near Land's End. Cable and Wireless also maintains a well-managed private archive at its London headquarters.

United States

Just as with Britain, the cable history of the United States rests on both corporate and government material. Unfortunately, many of the corporate records are unavailable. What is extant is exceedingly valuable. Foremost among this is the Western Union Collection at the Archives Center, National Museum of American History, Smithsonian Institution, Washington, D.C. These wide-ranging records include both corporate organizational documents and material of technical and cultural interest. AT&T, which dominated American cables after World War II, maintains a substantial archival collection in Warren, New Jersey (as a branch of the main Archives and Historic Center in San Antonio, Texas). Staff support is limited, but it is available to scholars by appointment. The records of the various cable companies besides Western Union—specifically those of the Mackay Companies and All America Cables—vanished into the corporate records of ITT and may be either with Alcatel, one of the ITT successor companies in international telecommunications, or may have been lost to historians altogether.

Absent a more complete corporate record, the historian needs to turn to government records, which are pleasingly vast. Materials at the National Archives and Records Administration (NARA) facilities in Washington, D.C., and College Park, Maryland, include the extensive holdings of the federal departments most interested in international cable communications. Among these are the Department of Commerce, State Department, Department of War, the U.S. Postal Service, and lesser agencies and commissions. These records cover the correspondence and subject files for the principal decision makers and offices interested in cable communications, and they quite frequently contain copies of correspondence and corporate information from the companies whose own records are now lost. Depending on the research issues, a good place to start might well be Record Group (RG) 40, the General Records of the Department of Commerce. In addition to senior department officials and the Office of the Secretary, components of the department that had an interest in cables included the Bureau of Foreign and Domestic Commerce (RG 151). The bureau's records from 1914 to 1958 contain a wealth of information not otherwise available on how federal officials worked with cable companies to push the creation of a domestic cable manufacturing industry.

Also important are the records of the State Department (RG 59). Any official actions by the United States government in international affairs goes through the State Department, even if the initiative for the policy comes from elsewhere. Researchers will want to make extensive use of the Central Decimal Files, covering the years 1910–1963. Because of the way in which the State Department established the decimal files, it is easy, after a few minutes' study, to locate the precise decimal reference for successive communications technologies (telephone, telegraph, cable, radio) in the records for each country or region of the world with which the United States had relations. Interest in cable communications by the U.S. military lay largely with the army, specifically the Signal Corps. The Office of the Chief Signal Officers (RG 111) regularly corresponded with companies on policy and research matters. This rich set of files should not be overlooked. Given the post office's fitful attention to telegraphic matters in the late nineteenth and early twentieth centuries, RG 28 should also not be overlooked. Because international conferences generated their own paperwork, these often rest in separate records groups. Some of these pertain to cable communications. One in particular is the Preliminary International Conference on Electrical Communications, held in Washington, D.C., in 1920. Records for this conference are part of the Records of International Conferences, Commissions and Expositions (RG 43). Additionally, the records of the Federal Communications Commission (RG 173), established in 1934, should supplement the records of the Department of State. Because of the nature of the archival system, however, other records may be located under their own numbers or form parts of other collections that might be of particular value but would not immediately come to the attention of a casual researcher. In any event, given the vast quantity of government documents available, plenty of rewards await those who search with patience and determination.

SOURCES IN DENMARK, RUSSIA, JAPAN, AND CHINA (DAQING YANG, WITH KURT JACOBSEN)

Denmark

Though not well known in the English-speaking world, the Great Northern Telegraph Company (GNTC; Det Store Nordiske Telegraf-Selskab) of Denmark, dating back to 1869, was once among the world's leading submarine telegraph companies. The company is no longer in the telecommunications business and has transferred its voluminous records to the Danish National Archives (Rigsarchiv) in Copenhagen. Its records are in Danish or English, due to its strong financial ties to London. The records also contain much information about Eastern and Associated Companies and the US-based Commercial Pacific Cable Company. The record's backbone is the very detailed and informative Directors Monthly Reports to the Board and the Board's Minutebooks. The records are open to researchers, but it is necessary to ask for permission in advance. The Foreign Office records and the Post Office records, kept in the National Archives, are also very informative because of the company's close cooperation with the Danish authorities.

With operations in Europe, Russia, and East Asia, some of GNTC's records remain overseas.

Russia

Because of the Great Northern Telegraph Company's strategic importance for Russian and Soviet international telecommunications, Russian archives contain many records related to the company and to international telegraphy in general. Pre-Soviet archives are kept in the Russian State Historical Archives in St. Petersburg. Information can be found in the Foreign Office records, Post Office records, and the Council of Ministers records. Archives from the Soviet period are kept in Moscow. The State Archives of the Russian Federation keeps the records from the Soviet government, while the Foreign Commissariat records are kept in the Foreign Policy Archives located in its own building in the Foreign Ministry of the Russian Federation. The lesser-known Economic Archive keeps the records from the Soviet Postal and Telegraphic Commissariat and the State Planning Commission, Gosplan. In principle all the records are open to researchers, but the recent changes in the political climate in Russia have made the working conditions in Russian archives rather uncertain. It is best to ask for permission in advance.

Japan

Japan emerged as a fledgling player in submarine telegraphy around the end of the nineteenth century. Many older records, however, seem to have been lost in the earthquake of 1923 and later in World War II. Telecommunications firms such as Nippon Telegraph and Telephone Co. (NTT) and KDDI, as successors to the prewar Ministry of Communications (MOC), which had held the monopoly since the 1880s, hold extensive archival materials on telecommunications operations. Since they are now private companies, public access to these documents is not guaranteed. The Communications Museum (Teishin Sogo Hakubutsukan, http://www.teipark.jp/e_about.html) in Tokyo has some original files dating from the prewar period that are open to the public. A researcher must also consider published sources. Though not primary documents, the 970-page *Kaiteisen hyakunen no ayumi* (A hundred years of submarine cables) published in 1971 by Nippon Denshin Denwa Kôsha Kaiteisen Shisetsu Jimusho is indispensable. It is based on an internally compiled collection of documents, *Nihon ni okeru kaiteisen* (Submarine cables in Japan) (1967), which I have not been able to locate.

China
Official documents related to submarine telegraphy and other aspects of telecommunications during the late Qing dynasty are located in No. 1 Archives in Beijing, China. Official documents for the Republican period, however, are scattered in Taiwan and mainland China. The Academia Sinica in Taiwan, which holds many of documents from Republic of China's Ministry of Foreign Affairs, published some of the materials on Japan-China negotiations over telecommunications in Zhongyang Yanjuyuan Jindaishi Yanjusuo, comp., *Zhong-Ri guanxi shiliao: Yudian hangyu yanlin jiaoshe, 1912–1916* (Historical sources of Sino-Japanese relations: Negotiations over post, telecommunications, navigation, fishery, salt, and forestry) (Taipei, 1975). Letters and telegrams concerning telecommunications in Qing China can be found in Wang Ermin, comp., *Sheng Xuanhuai shiye handian gao* (Letters and telegrams of Sheng Xuanhuai) (Taipei: Zhongyang Yanjiuyuan Jindaishi Yanjiusuo, 1993).

The People's Republic of China nationalized the Danish firm GNTC's operations in China in the early 1950s. As a result, over one thousand volumes of GNTC records are held at Shanghai, which served as the company's Far Eastern headquarters from 1871 to 1952. Housed in their original cabinets, they were open to researchers (as of early 2001), though permission from the Shanghai Telecommunications Administration (now part of the giant China Telecommunications Corporation) was needed. Many of these files have a brief Chinese summary, which seems to have been created in the late 1950s to identify those Chinese who had foreign connections.

OTHER PRINT SOURCES (PETER J. HUGILL)

The sources for a broad-based history of telegraphic, telephonic, wireless, and radar-based telecommunications are complex. In writing *Global Communications since 1844: Geopolitics and Technology* (Baltimore: Johns Hopkins University Press, 1999), I worked with many different sources.

The most useful broad-based sources of information about telecommunications were contemporary trade magazines, in particular the British weekly trade paper, *Electrician*, which began publication in 1861. At that time, academic journals were almost nonexistent, especially in engineering fields. *Electrician* covered the development of submarine telegraphy and the early history of wireless very well. It also performed such useful services as reprinting the entire testimony before the House of Commons Select Committee of what became known as the Marconi Scandal of 1912. I also used the monthly house publication of the Marconi Company, *Wireless World*, which began publication in April 1913, and its predecessor, *Marconigraph*, which is much harder to find. *Marconigraph* is the crucial source for understanding the Marconi Company's attempts to create the world's first global wireless system as an alternative to the submarine cables. The Marconi Scandal effectively ended that attempt. The only complete run of *Marconigraph* in the London area is at the library of the Institution of Engineering and Technology (IET, formerly the Institution of Electrical Engineers), on the Embankment. The IET library also has good runs of more recent technical journals, such as Philips Technical Review. The Marconi Company archival holdings were unavailable when I was researching *Global Communications*. They have since been given to Oxford University and can be found in the Bodleian Library. Information about the holdings of the IET Library and Archives can be found at http://www.theiet.org/library.

In recent years the IEE hosted a number of conferences on the history of telecommunications and has a very useful publications series that covers many aspects in detail. The conference publications on the history of wireless and radar include papers by engineers who were central to the development of particular technologies that can be treated as primary sources. For example, Wilhelm Runge, the head of radar development at Telefunken during World War II, presented a paper on the development of radar for the Luftwaffe at that company during the early 1940s. The books are usually written by engineers turned historians rather than by professional historians.

The final primary source worth mentioning is that of modern enthusiast literature. Such technologies as vacuum tubes and early wireless, television, and even radar sets have attracted significant numbers of collectors. Vacuum tubes and military wireless sets, especially for tanks, have devoted followings. These collectors frequently publish magazines that give significant information about such technologies, although they should be used with care. Even so, they often go well beyond the more conventional museum collections that are more typically consulted by scholars.

I expected to find primary source information about the pioneering military use of wireless in World War I's war in the air in the National Archives in London—like most of its users I still prefer the term Public Record Office (PRO). Unfortunately, basic circuit designs had 50-year secrecy notices on them, although they were published in *Wireless World* and other, more ephemeral, enthusiast magazines for radio amateurs shortly after the end of World War I. I also expected to find discussions in the records of the Cabinet meetings filed in the PRO pertaining to the origins of the radar chain in the mid-1930s—and was disappointed. These may have been D-noticed (meaning that they will never be released). Official documentation on the pioneering military use of wireless in Britain was, in other words, mediocre or missing.

Museum Perspectives (Bernard Finn)

Artifacts have traditionally been underutilized as source material, a situation easily confirmed by a perusal of the citations in this volume. This is, to my mind, primarily a consequence of the lack of a historiographic tradition (even among museum curators). At the very least the artifacts provide an emotional connection with people and events from the past. This is especially true of the cable station museums at Hearts Content in Newfoundland and at Orleans on Cape Cod. Cables also landed at Porthcurno, mentioned above. Instruments provide physical evidence of design traditions, of craftsmanship, and of the use of materials. Manipulating them can provide a measure of the skill required of the operator and perhaps of the tedium of his or her job.

Artifacts associated with submarine telegraphy can be found in many of the world's technical museums, especially telecommunications museums. Of special note are the Smithsonian's National Museum of American History and the Science Museum in London, as well as the Hearts Content, Orleans, and Porthcurno sites mentioned above.

Some station documents can be found at Orleans. Letter books and records from Hearts Content have been preserved in the Provincial Archives in St. Johns; microfilm copies are available in the Archives Center of the National Museum of American History.

An electronic version of an exhibit "The Underwater Web: Cabling the Seas" is available on the Smithsonian Institution Libraries Web site. New material is added from time to time, and as of this writing the intention is to continue to make it a resource for research in this area.

Bibliography

Ahvenainen, Jorma. *The European Cable Companies in South America before the First World War.* Helsinki: Finnish Academy of Science and Letters, 2004.

———. *The Far Eastern Telegraphs: The History of Telegraphic Communications between the Far East, Europe and America before the First World War.* Helsinki: Suomalinen Tiedeakatemia, 1981.

———. *The History of the Caribbean Telegraphs before the First World War.* Helsinki: Suomalinen Tiedeakatemia, 1996.

Baark, Erik. *Lightning Wires: The Telegraph and China's Technological Modernization 1860–1890.* Westport, Conn.: Greenwood Press, 1997.

Baglehole, Kenneth. *A Century of Service: A Brief History of Cable and Wireless Ltd., 1868–1968.* London, 1979.

Barty-King, Hugh. *Girdle Round the Earth: The Story of Cable and Wireless and Its Predecessors to Mark the Group's Jubilee, 1929–1979.* London: Heinemann, 1979.

Beauchamp, Ken. *History of Telegraphy.* London: Institution of Electrical Engineers, 2001.

Berthold, Victor M. *The History of the Telephone and Telegraph in Brazil, 1851–1921.* New York: American Telephone and Telegraph Company, 1922.

Blondheim, Menahem. *News over the Wires: The Telegraph and the Flow of Public Information in America, 1844–1897.* Cambridge, Mass.: Harvard University Press, 1994.

Boyce, Robert. "Canada and the Pacific Cable Controversy, 1923–1928: Forgotten Source of Imperial Alienation." *Journal of Imperial and Commonwealth History* 26, no.1 (January 1998): 72–92.

———. "Imperial Dreams and National Realities: Britain, Canada and the Struggle for a Pacific Telegraph Cable, 1879–1902." *English Historical Review* 115, no. 460 (January 2000): 39–70.

———. "Submarine Cables as a Factor in Britain's Ascendency as a World Power." In *Kommunikationsrevolutionen: Die neuen Medien des 16. und 19. Jahrunderts*, ed. Michael North, 81–99. Cologne: Böhlau, 1995.

Brett, John W. *On the Origin and Progress of the Oceanic Electric Telegraph.* London: W. S. Johnson, 1858.

Bright, Charles. *Imperial Telegraphic Communication.* London: P. S. King & Son, 1911.

———. *Submarine Telegraphs: Their History, Construction, and Working.* London: Crosby Lockwood, 1898.

Brown, F. J. *The Cable and Wireless Communications of the World.* 2nd ed. London: Sir Isaac Pitman & Sons, 1930.

Cable and Wireless Limited. *The Cable and Wireless Communications of the World: Some Lectures and Papers on the Subject, 1924–1939.* Cambridge: W. Heffer and Sons, 1939.

Cain, Robert J. "Telegraph Cables of the British Empire, 1850–1900." PhD dissertation, Duke University, 1971.

Carter, Samuel. *Cyrus Field: Man of Two Worlds.* New York: Putnam, 1968.

Chandler, Alfred D., and James Cortada, eds. *A Nation Transformed by Information: How Information Has Shaped the United States from Colonial Times to the Present.* New York: Oxford University Press, 2000.

Coates, Vary T., and Bernard Finn. *A Retrospective Technology Assessment: Submarine Telegraphy; The Transatlantic Cable of 1866.* San Francisco: San Francisco Press, 1979.

Codding, George A. *The International Telecommunication Union: An Experiment in International Cooperation.* Leiden: E. J. Brill, 1952.

de Cogan, Donard. "Dr. E. O. W. Whitehose and the 1858 Trans-Atlantic Cable." *History of Technology* 10 (1985): 1–15.

———. "International Cable Telegraph and the Austro-Hungarian Monarchy." *Blätter für Technikgeschichte* 50 (1988–1990): 7–11.

———. "Jim Graves and the Valentia Telegraph Station." *Electronics and Power* 30 (1984): 523–528.

Denny, Ludwell. *America Conquers Britain: A Record of War.* New York: Alfred A. Knopf, 1930.

Dibner, Bern. *The Atlantic Cable.* Norwalk: Burndy Library, 1959.

Fagen, M. D., ed. *A History of Engineering and Science in the Bell System: The Early Years (1875–1925).* Murray Hill, N.J.: Bell Telephone Laboratories, 1975.

Field, Henry M. *The Story of the Atlantic Telegraph.* New York: Scribner's Sons, 1892.

Finn, Bernard S., ed. *Development of Submarine Cable Communications.* 2 vols. New York: Arno Press, 1970.

———. "Growing Pains at the Crossroads of the World: A Submarine Cable Station in the 1870s." *IEEE Proceedings* 64 (1976): 1287–1292.

Gannett, Ezra Stiles. *The Atlantic Telegraph: A Discourse Delivered in the First Church, August 8, 1858.* Boston: Crosby, Nichols, 1858.

Griset, Pascal. "Technical Systems and Strategy: Interncontinental Telecommunications in the First Quarter of the Twentieth Century." In *The Governance of Large Technical Systems*, ed. Olivier Coutard, 58–72. New York: Routledge, 1999.

Haglund, H. H. "Atlantic Telegraph Cable Centennial." *Western Union Technical Review* 12 (1958): 81–102.

Haigh, K. R. *Cableships and Submarine Cables.* London: Adlard Coles, 1968.

Hall, Peter, and Paschal Preston. *The Carrier Wave: New Information Technology and the Geography of Innovation, 1846–2003.* London: Unwin Hyman, 1988.

Hanaoka, Kaoru. *Hundred Years of Submarine Cables and the Pacific.* Translated from Japanese by Great Northern Telegraph Company. Tokyo, 1969.

Harris, C. R. "The Persian Gulf Submarine Telegraph of 1864." *Geographical Journal* 135 (1969): 169–190.

Headrick, Daniel R. *The Invisible Weapon: Telecommunications and International Politics, 1851–1945.* New York: Oxford University Press, 1991.

———. *The Tentacles of Progress: Technology Transfer in the Age of Imperialism, 1850–1940.* New York: Oxford University Press, 1988.

Headrick, Daniel, and Pascal Griset. "Submarine Telegraph Cables: Business and Politics, 1838–1939." *Business History Review* 75 (Autumn 2001): 543–578.

Hecht, Jeff. *City of Light: The Story of Fiber Optics.* New York: Oxford University Press, 1999.

Hogan, Michael. *Informal Entente: The Private Structure of Cooperation in Anglo-American Diplomacy, 1918–1928.* Columbia: University of Missouri Press, 1977.

Hudson, Heather E. *Communication Satellites: Their Development and Impact.* New York: Free Press, 1990.

Hugill, Peter J. *Global Communications since 1844: Geopolitics and Technology.* Baltimore: Johns Hopkins University Press, 1999.

Hunt, Bruce. "Michael Faraday, Cable Telegraphy and the Rise of Field Theory." *History of Technology* 13 (1991): 1–19.

Jacobsen, Kurt. "Diplomacy and International Business: The Great Northern Telegraph Company 1869–1921." In *Business and Society: Entrepreneurs, Politics and Networks in a Historical Perspective*, ed. Anne-Marie Kuijlaars, Kim Prouddon, and Joop Visser. Rotterdam: CBG, 2000.

———. "The Great Northern Telegraph Company: A Danish Company in the Service of Globalisation since 1869." In *Between National Histories and Global History*, Historiallinen Arkisto 110:4, ed. Stein Tønnesson, Juahni Koponen, Niels Steensgaard, and Thommy Svensson, 179–197. Helsinki: Finska Historiska Samfundet, 1997.

———. "The Great Northern Telegraph-Company and Russia: 130 Years of Cooperation in the Light of the Big Politics." *History of the Motherland* 4 (2000): 44–54; 5 (2000): 58–69; and 6 (2000): 28–43 (in Russian).

———. "The Great Northern Telegraph Company and the British Empire 1869–1945." In *Britain and Denmark. Political, Economic and Cultural Relations in the 19th and 20th Centuries*, ed. Jørgen Sevaldsen with Bo Bjørke and Claus Bjørn. Copenhagen: Museum Tusculanum Press; University of Copenhagen, 2003.

———. "In Struggle for Control over the Far Eastern Telegraphs: The Great Northern Telegraph-Company and Japan 1870–1943." In Hubert Bonin, Christophe Bouneau, Ludovic Cailluet, Alexandre Ferndandez, and Silvia Marzagalli, eds., *Transnational Companies 19th–20th Centuries.* Paris: Éditions P.L.A.G.E., 2002.

[Japan.] Ministry of Communications. *Outline of the History of Telegraphs in Japan.* Tokyo, 1892.

Jaras, Thomas. "Promoters and Public Servants: The Role of the British Government in the Expansion of Submarine Telegraphy (1860–1870)." PhD dissertation, Georgetown University, 1976.

Kennedy, Paul M. "Imperial Cable Communications and Strategy, 1870–1914." *English Historical Review* 86 (1971): 728–752.

Kieve, Jeffrey. *The Electric Telegraph: A Social and Economic History*. Newton Abbot: David & Charles, 1974.

Kingsbury, H., and R. A. Goodman. "Methods and Equipment in Cable Telegraphy." *Journal of the American Institute of Electrical Engineers* 70 (1932): 477–521.

Koerper, Philip E. "The Pacific Cable: A Link in the Telegraphic Communications of the British Empire." PhD dissertation, University of Georgia, 1971.

Lawford, G. L., and L. R. Nicholson. *The Telcon Story, 1850–1950*. London: Telegraph Construction and Maintenance Co., 1960.

Lubrano, Annteresa. *The Telegraph: How Technology Innovation Caused Social Change*. New York: Garland, 1997.

Mackinder, Halford J. "The Great Trade Routes (Their Connection with the Organization of Industry, Commerce, and Finance)." *Journal of the Institute of Bankers* 21 (1900): 1–6, 137–155, 266–273.

Mance, Osborne. *International Telecommunications*. London: Oxford University Press, 1944.

Marvin, Carolyn. *When the Old Technologies Were New: Thinking about Electric Communication in the Late Nineteenth Century*. New York: Oxford University, 1988.

Matsumae, Shigeyoshi. *Study on the Long Distance Communication System by Non-Loaded Cable*. Tokyo: Corona, 1938.

McKercher, B. J. C., ed. *Anglo-American Relations in the 1920s: The Struggle for Supremacy*. Basingstoke: Macmillan, 1990.

Nickles, David Paull. *Under the Wire: How the Telegraph Changed Diplomacy*. Cambridge, Mass.: Harvard University Press, 2003.

Nippon Denshin Denwa Kôsha Kaiteisen Shisetsu Jimusho. *Kaiteisen hyakunen no ayumi* [A Hundred Years of the Submarine Cables]. Tokyo, 1971.

O'Neill, E. F., ed. *A History of Engineering & Science in the Bell System*, vol. 7, *Transmission Technology (1925–1975)*. Indianapolis: AT&T Bell Laboratories, 1985.

O'Rorke, Richard J., Jr. *1990 World's Submarine Telephone Cable Systems*. Washington, D.C.: Government Printing Office, 1991.

Parkin, George R. "The Geographical Unity of the British Empire." *Scottish Geographical Magazine* 10 (1894): 225–242.

Read, Donald. *The Power of News: The History of Reuters*. Oxford: Oxford University Press, 1992.

Runge, Peter K., and Patrick R. Trischitta, eds. *Undersea Lightwave Communications*. New York: IEEE Press, 1986.

Saxon, Mills J. *The Press and Communications of the Empire*. London: W. Collins Sons, 1924.

Schreiner, George A. *Cables and Wireless and Their Role in the Foreign Relations of the United States.* Boston: Stratford, 1924.

Scrymser, James A. *Personal Reminiscences of James A. Scrymser.* New York: privately printed, 1915.

Seeligmann, Th. *Indiarubber and Gutta Percha: A Complete Practical Treatise on Indiarubber and Gutta Percha in Their Historical, Botanical, Arboricultural, Mechanical, Chemical, and Electrical Aspects.* London: Scott, Greenwood, 1903.

Slack, Edward R. "The Effect of the U.S. International Telecommunications Industry on the United States Balance of Payments." DBA dissertation, George Washington University, 1970.

Slater, Ernest. *One Hundred Years. The Story of W. T. Henley's Telegraph Works Company.* London: Henley's Telegraph Works, 1937.

Sobel, Robert. *ITT: The Management of Opportunity.* New York: Times Books, 1982.

Sprague, William Buell. *A Sermon Addressed to the Second Presbyterian Congregation, Albany, on Sunday Morning, September 5, 1858 on the Completion of the Atlantic Telegraph.* Albany: Charles van Benthuysen, 1858.

Thomson, William. "On Practical Methods for Rapid Signalling by the Electric Telegraph." *Proceedings of the Royal Society of London* 8 (1856): 299–307.

———. "On the Forces Concerned in the Laying and Lifting of Deep-Sea Cables." *Proceedings of the Royal Society of Edinburgh* 5 (1866): 495–509.

———. "On the Theory of the Electric Telegraph." *Proceedings of the Royal Society of London* 7 (1855): 382–399.

Tribolet, Leslie Bennett. *The International Aspects of Electrical Communications in the Pacific Area.* Baltimore: Johns Hopkins University Press, 1928.

Twenty-Seventh Anniversary of the First Atlantic Cable: Mr. Cyrus W. Field's Banquet at the Star and Garter Hotel, Richmond, on August 5th, 1885. London: Gilbert & Rivington, 1886.

United Kingdom. Parliament. *Report of the Joint Committee Appointed by the Lords of the Committee of Privy Council for Trade and the Atlantic Telegraph Company to Inquire into the Construction of Submarine Telegraph Cables.* London, 1861.

Winkler, Jonathan Reed. *Nexus: Strategic Communications and American Security in World War I.* Cambridge, Mass.: Harvard University Press, 2008.

Winseck, Dwayne R., and Robert M. Pike. *Communication and Empire: Media, Markets, and Globalization, 1860–1930.* Durham: Duke University Press, 2007.

Yang, Daqing. "From Cooperation to Conflict: Ending Great Northern Telegraph Company in Japan." In *Transnational Companies 19th–20th Centuries*, ed. Hubert Bonin, Christophe Bouneau, Ludovic Cailluet, Alexandre Ferndandez, and Silvia Marzagalli. Paris: Éditions P.L.A.G.E., 2002.

———. *Technology of Empire: Telecommunications and Japanese Expansion, 1882–1945.* Cambridge, Mass.: Harvard University Asia Center, 2009.

Index